T0135669

Development and analysis of the processing of hybrid textiles into endless-fibre-reinforced thermoplastic composites

Vom Fachbereich Produktionstechnik

der

UNIVERSITÄT BREMEN

Zur Erlangung des Grades
Doktor-Ingenieur
genehmigte

Dissertation

von Maximilian Koerdt, M. Sc.

Erstgutachter: Univ.-Prof. Dr.-Ing. Axel S. Herrmann, Universität Bremen
Zweitgutachter: Univ.-Prof. Clemens Dransfeld, Technische Universiteit Delft

Tag der mündlichen Prüfung: 26.02.2020

Science-Report aus dem Faserinstitut Bremen

Hrsg.: Prof. Dr.-Ing. Axel S. Herrmann

ISSN 1611-3861

Bibliographic information published by Die Deutsche Bibliothek

Die Deutsche Bibliothek lists this publication in the Deutsche Nationalbibliografie; detailed bibliographic data is available in the Internet at http://dnb.ddb.de.

ISBN 978-3-8325-5060-8

Logos Verlag Berlin GmbH
Comeniushof, Gubener Str. 47,
10243 Berlin
Tel.: +49 030 42 85 10 90
Fax: +49 030 42 85 10 92
INTERNET: http://www.logos-verlag.de

Für Rieke + Greta + x

Danksagung

Eine umfassende Aufgabenstellung kann nur mit einer guten Mannschaft erfolgreich und effizient bearbeitet werden. Am Faserinstitut Bremen e.V. hatte ich das Glück, mit einer solchen zusammenarbeiten zu dürfen. Dafür und für die Betreuung meiner wissenschaftlichen Arbeit gilt mein besonderer Dank Herrn Prof. Dr. Axel S. Herrmann. Für die Zweitkorrektur und inspirierende Diskussionen bedanke ich mich bei Prof. Clemens Dransfeld. Ebenso gilt mein Dank meinen weiteren Prüfern Prof. Dr. Lucio Colombi Ciacchi, Dr. Andre Stieglitz, Nadine Gushurst und Mareike Woestmann.

Die angesprochene Mannschaft bestand aus allen Kollegen des FIBRE, insbesondere Peter Rödig und seiner Truppe, dem Laborteam um Herrn Dr. Ernö Németh, sowie den Kollegen aus den Abteilungen Fertigungstechnologien und Simulation. Für die durchgehenden fachlichen Diskussionen und Anregungen möchte ich mich bei Dr. Patrick Schiebel und Prof. Dr. Christian Brauner bedanken. Außerdem wären die PEEK-Fasern nur halb so schön geworden ohne den Einsatz von Lars Bostan und Daniel Weigel.

Für die organisatorische Betreuung seitens des Faserinstituts sowie der Universität Bremen bedanke ich mich bei Frau Corinna Gonzales und Susanne Baass. Ebenso gilt mein Dank den Partnern Karl Mayer Technische Textilien GmbH, Saertex GmbH und Victrex Europa GmbH für die materialseitige Unterstützung.

Der wichtigste Mannschaftsteil war zweifellos meine Familie, die mir immer den Rücken freigehalten hat. Dafür bedanke ich mich insbesondere bei meinen Eltern und Schwiegereltern, sowie bei Greta für die willkommene Abwechslung. Am allermeisten aber bedanke ich mich bei meiner Rieke.

Kurzfassung

Die vorliegende Arbeit trägt zu einer Beschleunigung des Herstellungsprozesses von Verbundkomponenten mit kohlenstofffaserverstärkten Hochleistungs-Thermoplasten bei. Eine umfassende Materialcharakterisierung von zwei Polyetheretherketon Polymeren und verschiedenen Konfigurationen von Hybridtextilien, bestehend aus Verstärkungs- und thermoplastischen Matrixfasern, liefert Eingangsparameter für ein Konsolidierungsmodell. Dieses Modell umfasst die Prozessschritte Erwärmen, Imprägnieren und Kühlen mit dem Schwerpunkt auf der wissensbasierten Beschreibung des Imprägnierverhaltens im Mesomaßstab.

Wesentliche Prozessparameter werden im Modellmaßstab untersucht, um die Entwicklung des Isoforming-Prozesses, einem auf Hybridtextilien zugeschnittenen Thermoformprozess, zu unterstützen. Durch den Prozess werden insbesondere die Schritte Heizen und Kühlen im Vergleich zu klassischen variothermen Ansätzen deutlich beschleunigt. Das Verfahren trägt dazu bei, das Potenzial der Drapierbarkeit von Hybridtextilien zu nutzen.

Eine anschließende Sensitivitätsanalyse validiert das Konsolidierungsmodell. Neben dem angewandten Druck erweist sich die textile Konfiguration für eine erfolgreiche Konsolidierung als wichtigster Faktor. Darüber hinaus liefert eine Korrelation zwischen Porosität und den resultierenden mechanischen Parametern Informationen über empfohlene Maximalwerte der Porosität in Laminaten. Schließlich gibt eine Prozessanalyse, die Online-Dickenmessung, im Laminat eingebettete faseroptische Sensoren und ein transparentes Presswerkzeug umfasst, Aufschluss über die Machbarkeit einer Online-Überwachung des Imprägniervorgangs.

Abstract

This thesis contributes to an acceleration of the manufacturing process of composite components from hybrid textiles, which contain reinforcement fibres and polymer matrix fibres from high-performance thermoplastics. Hybrid textiles distinguish in terms of drapeability compared to thermoplastic tapes. A comprehensive material characterisation of two polyetheretherketone polymers and different configurations of hybrid textiles deliver input parameters for a consolidation model. This model comprises the process steps of heating, impregnation and cooling, with a focus laying on the knowledge-based description of impregnation on meso-scale.

Relevant process parameters are investigated on model scale to support the development of the Isoforming process, which is a new developed thermoforming process tailored for hybrid textiles. During the process, the steps of heating and cooling are accelerated remarkably compared to classic variothermal approaches.

A subsequent sensitivity analysis validates the consolidation model. Beside the applied pressure and impregnation time, the textile configuration is a governing factor, which is analysed to recommend parameters for a successful consolidation process. Furthermore, a comparison between porosity and resulting mechanical parameters delivers information about the recommendable maximum values of void contents in laminates. Finally, a process analysis, which comprises online thickness assessment, laminate-embedded fibre-optic sensors and a transparent tooling, provides information about the feasibility of an online assessment of the state of impregnation.

Table of contents

Abbreviations

CF	carbon fibre
CFRP	carbon-fibre-reinforced plastics
CT	computer tomography
CTE	coefficient of thermal expansion
DMA	dynamic mechanical thermal analysis
DoC	degree of crystallisation
DSC	differential scanning calorimetry
FEA	finite element analysis
FOS	fibre-optic sensors
GF	glass fibre
ILSS	inter-laminar shear strength
IR	infrared
LFA	laser flash analysis
NCF	non-crimp fabric
NELF	non-equilibrium lattice fluid theory
NLR	National Aerospace Laboratory
ODE	ordinary differential equation
PA	polyamide
PEAK	polyaryletherketone
PEEK	polyetheretherketone
PEKK	polyetherketoneketone
PP	polypropylene
PPS	polyphenylene sulphide
PVA	polyvinyl alcohol
RIM	reaction injection moulding
RTM	resin transfer moulding
STD	standard deviation

TFP	tailored fibre placement
TMA	thermomechanical analysis
TMDSC	temperature-modulated differential scanning calorimeter
TP	thermoplastic
WLF	Williams–Landel–Ferry

Notation

Symbol	Unit	Description
a_T	mm/s^2	thermal diffusivity
b, k, l, m, n	1	empirical parameters
$B(T)$	1	empirical parameter for Avrami model
CTE	1/K	coefficient of thermal expansion
c_p	J/[g K]	specific heat capacity
D_h	1	degree of healing
E^*	MPa	complex modulus
E_0	J/mol	material-specific activation energy
ΔH_m	J/g	enthalpy of fusion
ΔH_m^0	J/g	enthalpy of fusion for entire crystalline spherulite
K	m^2	permeability
k_z	1	permeability constant
L	m	impregnated length
$L_{shear-edge}$	m	shear-edge length
n	1	number of moles per molecule
p	MPa	pressure
p_G	MPa	pressure inside entrapped air bubble inside voids
p_M	MPa	pressure inside the matrix system
p_{app}	MPa	applied pressure
p_{cap}	MPa	capillary pressure
p_{max}	MPa	pressure required to compress to φ_{Fmax}
$\dot{q}$	J/s	heat flow
R	J/[mol K]	gas constant

Re	1	Reynolds number
r_f	μm	filament diameter
s	1	non-dimensional position of fibre front
T	°C	temperature
T_0	°C	reference temperature
T_g	°C	glass transition temperature
T_m	°C	melting temperature
$t_{1/2}$	s	crystallisation half time in Nakamura model
t_r	s	reptation times
th_{CF}	m	initial thickness of single carbon-fibre ply
th	m	thickness
$th_{shear-edge}$	m	shear-edge thickness
u_i	m/s	velocity
u_F	m/s	velocity of fibre front
u_M	m/s	velocity of matrix front
Vol	m^3	volume
$\dot{V}$	m^3/s	volume flow
vc	1	void content
$w_{shear-edge}$	m	shear-edge width
$X_c(t)$	1	absolute crystallinity at time t
$X_{c\infty}$	1	heat of fusion resp. max. absolute crystallinity
x, y, z	1	dimensions in artesian coordinates
z_c	m	critical impregnation length for fully entrapped air
z_e	m	interface between neat matrix and impregnated
z_f	m	position of the infiltration front

β	1	parameter in Boltzmann transformation
γ_{se}	N/m	surface energy
$\dot{\gamma}$	Hz	shear rate
η	Pa s	dynamic viscosity
η_0	Pa s	kinematic viscosity at reference temperature
θ	°	contact angle
κ	1	rel. critical impregnation length for entrapped air
κ_F	1	elastic constant of the fibre bed

λ_T	W/[m K]	thermal conductivity
μ	m²/s	kinematic viscosity
ρ	g/cm³	density
σ_F	MPa	pressure acting on textile preform
ς	1	Carman-Kozeny constant
φ_F	1	fibre volume content
φ_{F0}	1	initial FVC at first compressive load
$\varphi_{F,max}$	1	maximum fibre volume content
φ_{Fr}	1	FVC in relaxed state after first compression
φ_{max}	1	maximum available fibre volume fraction
φ_{micro}	1	fibre volume content on micro-scale
χ	1	non-dimensional position in impregnated region
Ψ	1	non-dimensional flow rate

Subscripts

$\perp$	transversal
$\parallel$	longitudinal
M	matrix
F	fibre
lam	laminate
max	maximum
x, y, z	cartesian coordinates
c	compressed state
CF	carbon fibre
$micro$	micro-scale
cap	capillary
r	relaxed state
$impr$	impregnated state

1 Introduction

High-performance carbon-fibre-reinforced composite materials are widely used in aerospace structures as they offer many advantages: they are lightweight, have high specific stiffness as well as strength and are very durable. In order to attain the full potential of composite material, the disadvantages like the relatively high material and manufacturing costs must be weighed against the advantages. To meet the needs of emerging markets, the technologies for processing carbon-fibre-reinforced plastics (CFRP) require a significant cost reduction. Especially, complex structural components in aviation applications are often produced using autoclaving process or resin transfer moulding (RTM) process, since the processed reinforcement fibres offer the great potential of being drapable into complex geometries. However, long cycle times reduce the economic process efficiency remarkably.

A promising strategy to decrease cycle times is processing of thermoplastic composites due to their fast processability, since no cross-linking of molecular chains is required as for thermoset resin systems. Nevertheless, nowadays thermoplastic CFRP are predominantly manufactured with pre-impregnated laminates, which results in limited freedom of design and drapeability. Hybrid textiles consisting of thermoplastic fibres and carbon fibres can obviate this disadvantage, since they combine the drapeability of dry textiles with thermoplastic matrices, which furthermore allows net-shape components. On the one hand, relative shifting between the fibres, and consequently draping, still is possible in a preforming step. On the other hand, due to viscosities, which are three decades higher than for thermoset resins, the impregnation slows down significantly. Regarding the increasing demand for lightweight solutions, thermoplastic composites also provide effective solutions for comprehensive recycling. Due to possible re-forming, the feasibility of recycling is considerably improved compared to thermoset composites.

The success of thermoplastic composites in aircraft applications will be decisively depending on promising compromises between short times for heating, impregnation and cooling, costs for material and requirements according to the specific mechanical properties of the components manufactured. This thesis shall contribute to develop a processing strategy tailored to the distinct requirements of hybrid textiles for composite structures which satisfy the high demands of the aircraft industry.

1.1 Objectives

A fundamental prerequisite to develop sustainable thermoplastic composites for aircraft purposes is to pursue a holistic approach capturing material and process development together. This thesis thus shall contribute to obtain a better understanding about the influence of material and process parameters on the structure of the resulting thermoplastic composite part, namely the polymer raw materials, so as different textile configurations will be investigated. A process model will derive adequate process conditions for consolidation, while consolidation experiments will help to deliver input parameters for the model. The development of an efficient consolidation process shall finally show the potential of a fast and high-quality manufacturing process for high-performance thermoplastic composite structures. Concluding, the objectives of this thesis concerning the different topics are as follows:

Material characterisation

- Material analysis of low-melt PEAK and comparison to classic PEEK to give recommendations for design and process parameters
- Characterisation of four different textile configurations to give representative information about the influence of the textile topology on consolidation

Process modelling

- Develop impregnation model for first-time-right processing

Process development

- Development of fast and cost-efficient compression moulding process designed for hybrid textiles for near-net-shape composite parts

Process analysis

- Investigation of the influence of laminate porosity
- Analysis of consolidation of hybrid textiles to thermoplastic composites to give recommendations for process design and modelling

1.2 Scope of work

In order to achieve optimised laminate properties, the knowledge-based development requires a prior investigation of the distinctive influencing phenomena. In this context, a material characterisation delivers appropriate data for an impregnation model. This model helps in finding process parameters for optimising the developed compression moulding process. The framework also comprises process monitoring, using in situ measurement of the laminate thickness, on the one hand and the aggregate state of the polymer, using fibre-optic sensors, on the other hand. Furthermore, a visual insight into macro-impregnation is given. In this context, the steps of model development, process development and process analysis interact with each other, so that a sharp chronological distinction is not feasible. In summary, the scope of work is illustrated below (Figure 1).

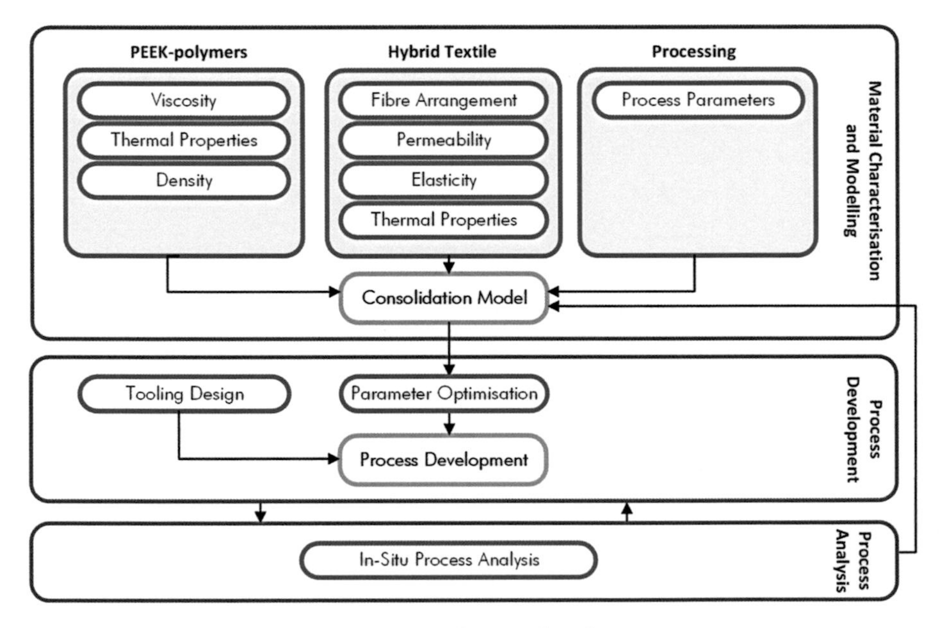

Figure 1 – Scope of work

2 State of the art

This introducing chapter provides an overview about the currently applicable raw materials for thermoplastic composites for aerospace primary load-carrying structures. Furthermore, existing manufacturing processes are briefly introduced.

In recent years, thermoplastic composites became more and more interesting for future aircraft applications due to their advantageous processing properties. The main advantage of this material class is out-of-autoclave processability and thus fast cycle times. Especially, semi-crystalline thermoplastics are of interest due to their mechanical properties and improved chemical resistance. Examples of current thermoplastic components made from polyetheretherketone (PEEK) already in service or in development are shown in Figure 2.

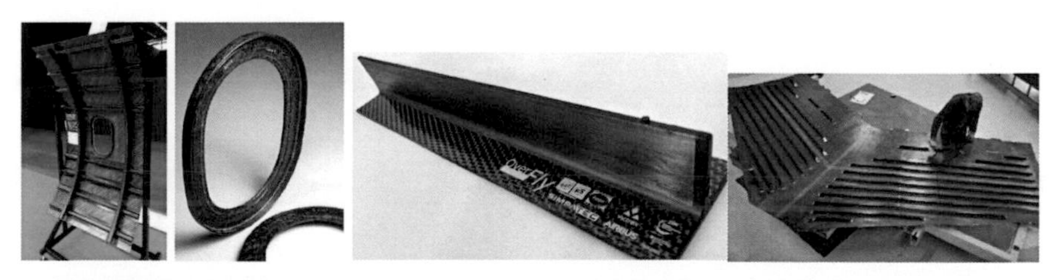

Figure 2 – Exemplary aircraft structures made from thermoplastic composites in autoclave cycles, compression moulding setup, overmoulding or welding processes (f.l.t.r.: Eco-design demonstrator [CleanSky-project], thermoplastic window frame [NLR], overmoulding demonstrator [ZIM-project OverFly] and Gulfstream G650 Tail [Fokker])

Materials and phenomena

Based on the scale of arrangement of the two semi-finished components (reinforcement fibres and matrix fibres), hybrid thermoplastic carbon-fibre-reinforced materials can be classified into three types. They are hybrid rovings, for example, commingled yarns (mix of reinforcement fibres and matrix fibres in the roving); hybrid textiles, for example, unmingled yarns (non-crimp fabric [NCF] or fabric of reinforcement fibres and matrix fibres); and hybrid laminates from film-stacking processes. Beside hybrid laminates, which are preferably used for pre-impregnation of thermoplastic laminates in a film-stacking arrangement, hybrid rovings and hybrid textiles are limited to very special applications.

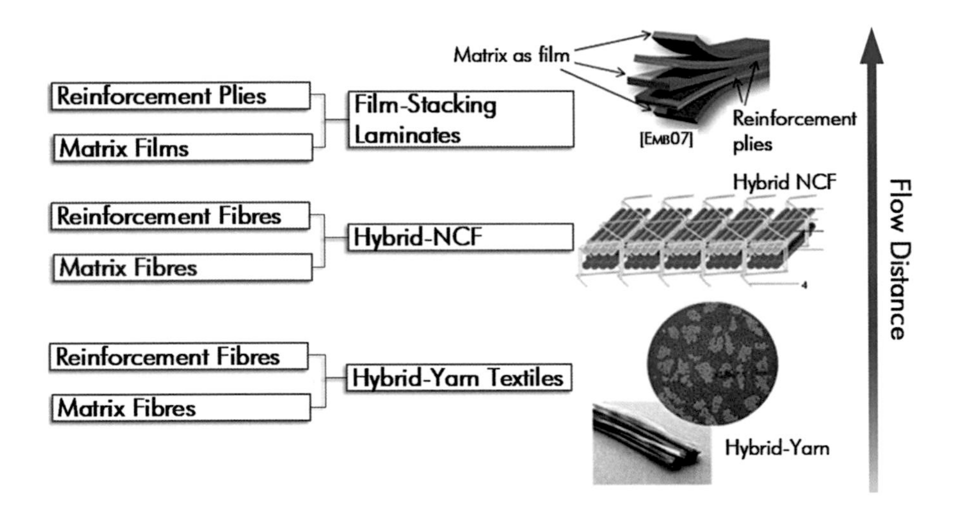

Figure 3 – Classification of hybrid TP-CF systems

These introduced thermoplastic-fibre and reinforcement-fibre systems offer the possibility to reduce raw material costs concerning pre-impregnated tapes since impregnation and consolidation are done at the same time in the manufacturing process (not before). Furthermore, the in-plane drapability of hybrid textiles allows a significantly higher freedom of design compared to pre-impregnated materials, which can also amplify the lightweight aspect.

Because of viscosities, which are about three decades higher for thermoplastic polymers as PEEK than for convenient thermoset resin systems, it is a key requirement for impregnation to minimise flow distances. A higher degree of mixing, as shown in Figure 3, therefore simplifies the impregnation process of hybrid textiles remarkably.

In order to achieve the required laminate quality, a knowledge-based understanding about the consolidation process is necessary, which can be divided into the following sub-processes: fibre bed compaction and flow, impregnation and ply compaction, and redistribution of the polymer fluid. These processes appear simultaneously [WYS07,

ROU13]. Three components, namely fibre, matrix and voids, will interact on macro-, meso- and micro-scales. While the macro-scale comprises several adjacent rovings in several plies, the meso-scale is defined by the environment of one roving, and the micro-scale is denoted by the region of adjacent filaments.

On the meso-scale, the so-called inter-ply deformation will be affected by external forces and deformations and lead to laminate deformation, fibre reorientation and seepage flow respective wet out. On the micro-scale, the so-called intra-ply deformation will lead to ply deformation and infiltration flow affected by fluid pressure and ply compression. Key parameter for the impregnation, which is also described by Darcy´s law, is the flow length.

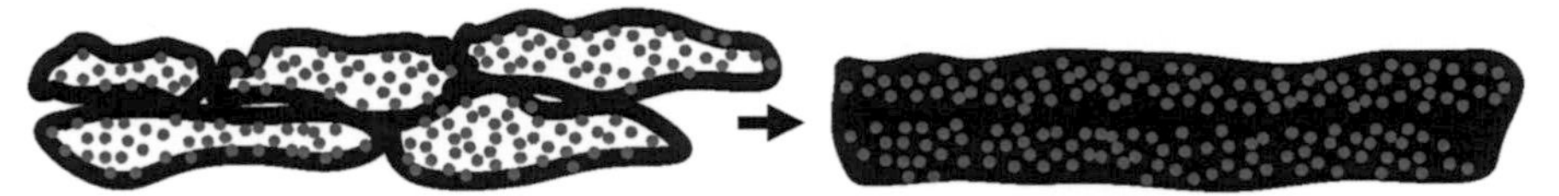

Figure 4 – Impregnation process [acc. to WYS09]

All these individual effects have been the object of investigation in the past. The deformation of the fibre bed during consolidation is driven by forces resulting from pressing and redistribution of the matrix fluid. Deformation takes place on laminate level by interply shear and slip, on intraply level by intraply shear and on fibre level by fibre motion in and transverse to the fibre direction.

Since impregnation is highly dependent on ply compaction, both phenomena are often observed in parallel. Different scientific clusters discussed the corresponding effects. For ply compaction, an unsaturated compaction of the plies was assumed in the majority of the cases. Various investigations were performed varying the ply orientations, areal weight of single plies and, of course, the fibre volume content as governing parameters [KIM91, TOL98]. Rouhi and Wysocki analysed the effect of relaxation of the fibre bed with respect to time and initial degree of compression and they obtained the resulting conclusion that relaxation can reduce internal pressure forces by more than 50%, and thus, the effect should be considered [ROU15].

For modelling the impregnation today, still the model of Darcy is generally used. Several projects successfully applied Darcy´s law for thermoplastic impregnation of reinforcement textiles [SCO10, JES08, BER01, MIC00, STU16, BOU00]. All models have the problem of investigating permeability in common, which can be approached by simple analytic models as the Gebart model [GEB91] or by experimental analysis taking material effects like sizing into account, as for the Carman–Kozeny model with its experimentally investigated coefficient of permeability [GUT97].

Thermoforming processes

In order to give an understanding of the consolidation process, its main phases are illustrated in Figure 5. The process begins by heating the raw material to processing temperature, which is above the melting temperature of the polymer. At processing temperature, pressure is applied to provoke impregnation of the polymer fluid in-between the filaments of the rovings. After impregnation, the setup is cooled below the glass transition temperature of the polymer. In this cooling phase, crystallisation of the polymer takes place, which directly effects material properties as, for instance, chemical resistance or process-induced shrinkage.

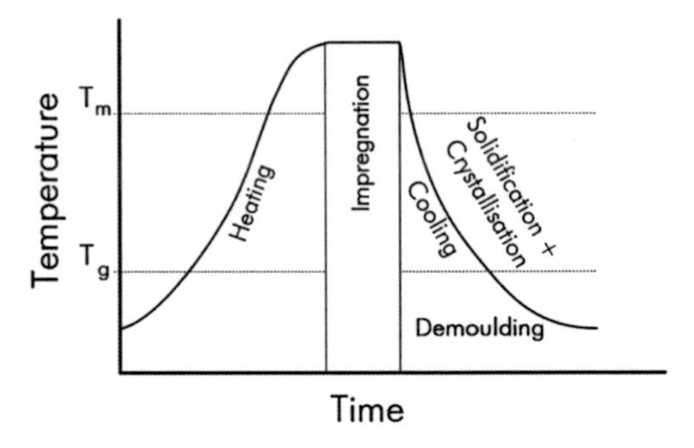

Figure 5 – Typical consolidation-process for hybrid TP-CF systems (melting, consolidation and solidification) [acc. to BOU01]

Today, numerous processes are applicable for manufacturing structural thermoplastic composites. A selection for thermoplastic composites is explained in the following.

Nowadays, most structural thermoplastic composite parts for aircraft applications are manufactured by thermoforming of pre-consolidated thermoplastic laminates. The laminate is externally heated above the melting temperature by IR radiation for instance, transferred into the tooling mould and subsequently pressed and cooled down by the isothermal tooling. This process is beneficial for high production rates due to its possible cycle times in the range of a minute and high industrial standard. Furthermore, complex shaped parts can be manufactured, mainly limited by the drapeability of the textile. Recently, the process is combined with injection moulding, so that even more complex geometries are feasible [SCU11]. A constraint of classic thermoforming for aircraft structures is its limitation to preforms that require a certain amount of internal cohesion, which comes from interlocked fibres in fabrics or the molten thermoplastic polymer. This restraint also effects that in-plane draping is not applicable in classical thermoforming. An overview about the dependencies of mechanical performance and flexibility in design is given in Figure 6.

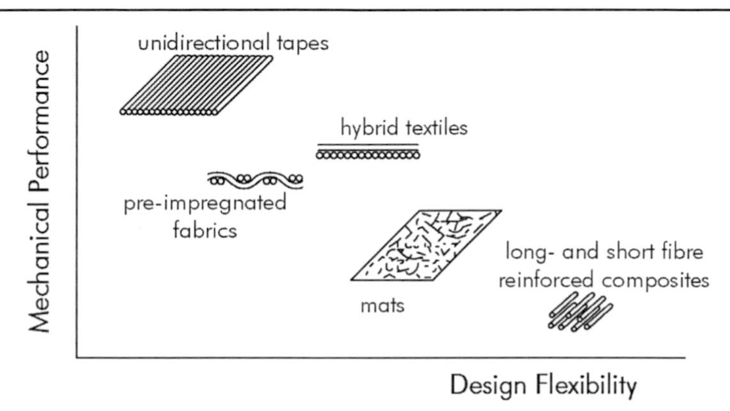

Figure 6 – Comparison of fibre reinforcements

As for large thermoset parts, the use of an autoclave is also applicable to thermoplastic composites [MAN89]. Both pre-impregnated fibres and hybrid textiles can be processed in this way. However, the process is very challenging for thermoplastic composites due to several aspects like a limited pressure for autoclaves (usually max. 1 MPa). Temperatures up to 400°C pose the most relevant problem in terms of thermal stability of the manufacturing equipment (vacuum bags, sealant tapes, breather fabric, etc.) and in terms of thermal expansion of the tools and jigs.

Press forming is a common process for impregnating reinforcement textiles with thermoplastic matrix. In this context, textile and matrix in films or fibres are deposited inside a mould or a flat tooling, which is heated actively to melt the thermoplastic material. During the pressure stage, impregnation takes place and finally the whole setting is cooled, so that subsequently the laminate can be removed from the tooling. A common approach is heating of the tooling by oil or electric heating elements and cooling by oil or water [KOE16]. Again, high temperatures up to 400°C require extensive efforts. Especially, regular thermal oil specifications necessitate an external pressure, which increases precautions regarding safety and thus costs.

Beside classic variothermal heating of a compression mould, numerous variations exist. High process temperatures and a high melt viscosity are usually the problems to face. Feasible solutions are found in fast variothermal processes with heating/cooling channels approximately 10 mm below the complex shaped cavity surface, combined with a minimised cavity weight to reduce thermal masses. Temperature gradients up to 7–10 K/s were attained for the processing of glass-fibre-reinforced PA6 [HOP14]. As heating element inside press toolings, the use of ceramic material silicon carbide (SiC) is another approach. It is advantageous compared to cartridge-heated steel toolings with respect to its higher thermal conductivity, its smaller thermal expansion and its reduced heat capacity. Accordingly, high heating rates up to 70 K/min to 190°C were reached with respect to 37 K/min for a classic steel tooling. Major disadvantage of this concept is the brittle behaviour of ceramics, which demands a ceramic-appropriate tooling design [KUE16].

A further approach is the use of induction technology where either the tooling or susceptors integrated in thermoplastic fibres are heated by an electromagnetic field [BOS14]. The first-mentioned technology is based on induction coils integrated into the tooling, which create the electromagnetic field that is optimised regarding the shape of the cavity to heat it locally. Fast heating rates up to 400°C within a few minutes are possible. Cooling is accomplished by water, which is purged by air before heating. The process is already industrialised and stable, thus giving a good alternative for consolidating high-temperature hybrid textiles [SCA17]. Nevertheless, lot sizes need to be high to regain the high invests for the tooling including necessary equipment. Direct inductive heating of thermoplastic fibres requires integration of susceptors into the fibres. These magnetisable particles in scales of a few micrometres heat up inside an electromagnetic field and can thus heat up the polymer within seconds to 400°C. This process still is in the phase of development. A major challenge is the realisation of a high electromagnetic field strength close to the fibres [SCH18].

Another method to perform consolidation is the conjunction of press forming with injection moulding. Currently, reaction injection moulding (RIM) processes enable the impregnation of a textile with thermoplastic monomers, which have the advantage of low viscosities in ranges of 5–90 Pa s, consequently also being processable by in-plane injection [KM16, RIJ07]. After impregnation, polycondensation takes place inside the mould. Nowadays, this process has been industrialised only for polymers based on polyamide and polyurethane. To avoid this drawback, a direct through-thickness thermoplastic melt impregnation by injection moulding has been investigated by Studer et al. [STU16]. For impregnation, the thermoplastic melt is filled into a gap above the preform inside the shear-edge tooling. After the filling of the gap, pressure is applied to the melt and impregnation of the preform is provoked. Despite the advantage that all polymers which are injectable can be used, it is necessary to consider only polymers with low viscosities, since the complete stacking of the preform must be penetrated.

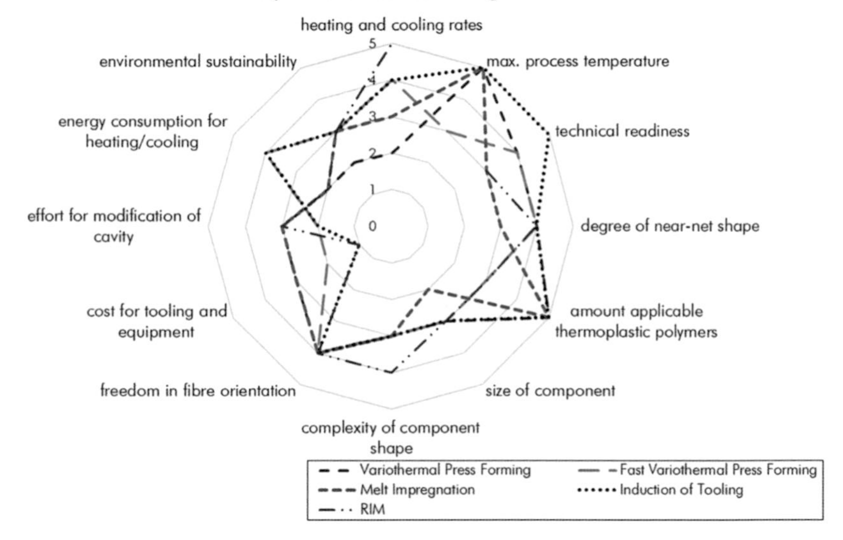

Figure 7 – Comparison of several thermoforming processes (high value favourable)

In summary, various processes for consolidating thermoplastic composites are applicable. All of them interact in terms of costs for toolings and equipment, variability concerning the choice of the polymer, variability concerning the shape and the size of the specimen, maximum temperature, heating and cooling rates, or technical readiness.

The most interesting processes for medium size parts are compared in Figure 7. The illustration shows that there still is a demand for consolidation processes, which enable cost-effective adjusting of the component shape for fast processes, especially fast and energy-efficient heating and cooling.

3 Process-dependent material behaviour – PEEK polymers

In the subsequent chapter, two polyaryletherketone (PAEK)-based polymers are characterised regarding all relevant material parameters to understand or model the process behaviour during compression moulding with fibre reinforcements. PAEK and, especially, their polyetheretherketone variants are widely spread for applications in aviation, medicine or petro-industry. In particular, 151G and AE250™-polymer, provided by VICTREX, are examined. Advantages are, for instance, their high thermal stability up to 280°C, their sterilisability and their resistance against numerous chemical products.

While 151G is commercially available and known for decades, AE250™-polymer represents a new polymer, especially applicable for overmoulding components. Its melting temperature is about 40 K below the melting temperature of classic PEEK used for injection moulding, which especially is advantageous for overmoulding processes to create a sustainable fusion of an endless-fibre-reinforced insert based on low-melt PEEK and an injection-moulded classic PEEK.

Both materials are analysed and compared concerning their

- viscosity
- crystallisation behaviour during cooling
- specific heat capacity
- thermal conductivity
- coefficient of thermal expansion
- temperature-dependent flexural modulus (only 151G)
- temperature-dependent density.

3.1 Viscosity

During composite processing, the viscosity directly influences the impregnation behaviour. In order to improve impregnation, the viscosities shall be reduced to a minimum.

The resistance of a fluid to motion deformation by shear stresses is described by the viscosity. Parameters affecting the viscosity of a thermoplastic melt are its temperature and the shear stresses acting on the flow. High temperatures generally reduce the viscosity due to an increased distance between the polymer molecules, which is preferable for impregnation. PEEK polymers belong to the group of shear-thinning fluids, which means that higher shear rates reduce the resistance against deformation and thus viscosity. This phenomenon can be explained by rearrangement of molecular chains during deformation. As the rate of deformation grows, the corresponding increasing shear stresses disentangle the molecules. The molecular chains are thus able to slip against each other into the direction of deformation resulting in a reduction of viscosity.

For compression moulding of composites containing PEEK as matrix system, processing temperatures shall be in the range between 360°C and 420°C. This assures a completely molten polymer on the one side and prevents thermal degradation of the polymer on the other hand. For compression moulding of endless-fibre-reinforced thermoplastics, the low flow rates lead to low shear rates below 1 s^{-1}, as indicated by several authors [ADV10, OSW15].

Literature and models

Appropriate mathematical models to predict the shear rate-dependent viscosity of thermoplastic materials are the Arrhenius model, the Williams–Landel–Ferry (WLF) model or the Carreau model.

The Arrhenius model allows a prediction of the viscosity at given temperature considering the shear rate $\dot{\gamma}$ for viscous flow as illustrated in equations (3.1) and (3.2) [ARH16]:

$$\eta(T) = \eta_0(T_0) \cdot e^{\frac{E_0}{R}\left(\frac{1}{T}-\frac{1}{T_0}\right)}. \tag{3.1}$$

The model requires the material-specific activation energy E_0, which is influenced by the shear rate. It can be evaluated experimentally. Therefore, equation (3.1) is rewritten to:

$$\eta(T,\dot{\gamma}) = \eta_0(T_0) \cdot e^{\frac{E_0(\dot{\gamma})}{R}\left(\frac{1}{T}-\frac{1}{T_0}\right)}. \quad (3.2)$$

Since the Arrhenius model fits the experimental data appropriately, it is used to express the temperature-dependent and shear rate-dependent viscosity. Rearrangement of equation (3.2) delivers the data for the shear rate-dependent activation energy in equation (3.3):

$$E_0(\dot{\gamma}) = \frac{R}{\left(\frac{1}{T}-\frac{1}{T_0}\right)} \cdot ln\left(\frac{\eta(T,\dot{\gamma})}{\eta(T_0,\dot{\gamma})}\right). \quad (3.3)$$

With R as gas constant taken from Arrhenius and measured data for viscosity at discrete temperatures, values for $E_0(\dot{\gamma})$ can be identified and subsequently interpolated for varying shear rates [ARH16].

Parameters and results

In order to determine the temperature-dependent viscosity behaviour, a parallel-plate rheometer AR2000 by TA Instruments was used. The investigation of the temperature dependency was performed by heating the specimen after a previous melting and solidification cycle to assure a reproducible melting point and to eliminate residual moisture. For measuring, the following parameters were chosen (Table 1):

Figure 8 – PEEK specimen in molten state between rheometer plates

Table 1 – Parameters for the measurement of viscosity

		AE250™-polymer	151G
Oscillation rate	[Hz]	2,5; 5; 10	
Temperature range	[°C]	360 - 400	310 - 360
Displacement	[rad]	8e-04	
Heating rate	[K/min]	5	5
Gap	[µm]	1000	
Specimen diameter	[mm]	25	
Repetitions	[1]	1	

During the measurement, data were logged continuously. Averaged values are listed in Tables 2 and Table 3. For both polymers, the standard deviation of the measured values for viscosity was in a satisfactory range below 2%. Especially for small oscillation rates of 2,5 Hz to 5 Hz, the experiments provide comparable values. Thus, the dependency between the shear rate and viscosity can be neglected for shear rates below 5 Hz for the analysed polymers.

Table 2 – Measured viscosity for PEEK AE250™-polymer

Osc. rate [Hz]	Viscosity [Pa s]				$\overline{E_0}(\dot{\gamma})$	STD
	320°C	330°C	340°C	350°C	[J/mol]	[%]
2,5	401	342	295	255	14095	2
5	403	335	383	241	16337	2
10	346	295	254	221	14043	1

Table 3 – Measured viscosity for PEEK 151G

Osc. rate [Hz]	Viscosity [Pa s]					$\overline{E_0}(\dot{\gamma})$	STD
	360°C	370°C	380°C	390°C	400°C	[J/mol]	[%]
2,5	386	339	298	264	236	13047	5
5	375	342	312	285	259	10401	3
10	287	256	233	211	198	10979	3

note: $\overline{E_0}(2{,}5\ Hz)$ only based on values from 370-400°C

Modelling

The determined data illustrate an almost constant activation energy for constant oscillation rates at different temperatures. From this correlation, the shear rate-dependent and temperature-dependent viscosity for both PEEK polymers result in the following parameters for the Arrhenius model (Table 4):

Table 4 – Parameters for modelling the viscosity

	PEEK 151G			PEEK AE250™-polymer		
Osc. rate [Hz]	η_0	T_0	$\overline{E_0}(\dot{\gamma})$	η_0	T_0	$\overline{E_0}(\dot{\gamma})$
2,5	386	360°C	13047	401	320°C	14095
5	375	360°C	10401	403	320°C	16337
10	287	360°C	10979	346	320°C	14043
Gas constant [J / mol K]		8,314		[ARH16]		

Comparing the measured data with the Arrhenius model shows a good correlation as shown in Figure 9. Consequently, it will be considered for modelling viscosities for modelling impregnation in chapter 6, assuming low shear rates of 2,5 Hz.

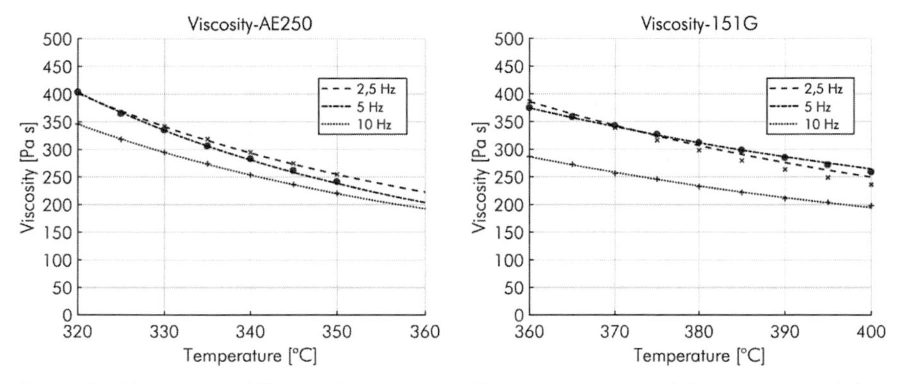

Figure 9 –Viscosity at different shear rates and temperatures (model and averaged data points)

Conclusion

To conclude, the viscosities of both polymers provide comparable values for shear rates up to 5 Hz oscillation rate. At processing temperatures of 400°C, respectively 40 K below at 360°C for AE250™-polymer, a viscosity between 200 and 250 Pa s can be expected.

3.2 Crystallinity

Thermoplastics can be separated into semi-crystalline and amorphous structures. Since most of the high-performance technical thermoplastics (like PEEK, PPS and PA) are semi-crystalline, this chapter will focus on them. Semi-crystalline polymers are characterised by containing a fraction of crystalline-orientated molecules and amorphous regions (Greek: 'without form'). During solidification and the simultaneous transition phase from molten to the glass transition temperature, the molecules are free to move and align themselves partially into a crystalline structure. The degree of crystallisation of a semi-crystalline polymer depends on the cooling rate, melting temperature, chemical configuration, molecular weight, shear flow, pressure and possible impurities as fibres, for instance.

During the first phase of crystallisation of a fibre-reinforced semi-crystalline polymer, nucleation begins initiated by filaments which act as impurity. Note that this is only valid for small shear rates, as they appear during compression moulding. For high shear rates, which, for instance, appear during injection moulding, also a flow-induced crystallisation can occur. The second phase of crystallisation is based on growing crystalline regions. After the subsequent solidification, the post-crystallisation phase is a long-term effect which results from rearrangement of molecules at temperatures below T_g. Post-crystallisation increases internal stresses which can also provoke deformations after part removal.

Influences of DoC

The knowledge about the crystallisation behaviour of the processed polymers is important since crystallisation is a key parameter for several material properties. The degree of crystallisation (DoC) is directly influencing the Young's modulus, the resistance against environmental influences, the density, the coefficient of thermal expansion (CTE) or the thermal shrinkage respective process-induced deformations. This chapter gives an overview about the crystallisation characteristic to comprise a better understanding of the consolidation process during compression moulding of endless-fibre-reinforced thermoplastics.

Internal stresses and warpage

Wijskamp assumed that from the crystallisation half time the point of solidification is crossed during cooling. From this point on, thermal or crystallisation-induced shrinkage is modelled to cause deformations respective internal stresses [Wij05]. This assumption of course does not respect that the polymer network can transfer stresses already previously. However, crystalline structures are mechanically superior to amorphous regions, which supports the simplification, that from this so-called solidification point, significant shrinkage-induced stresses arise. The solidification point is close to the maximum of the exothermal heat flux caused by crystallisation and can thus be approximated by analysing the measured heat flux during cooling. Normally, the corresponding temperature is close to the peak crystallisation temperature which is highly dependent on the cooling rate, the shear flow or the crystallisation kinetics.

For modelling the crystallisation kinetics, the Avrami equation states a suitable approach for an isothermal crystallisation:

$$DoC = \frac{X_c(t)}{X_{c\infty}} = 1 - e^{-B(T)t^n}.$$

Here $X_c(t)$ represents the absolute crystallinity at time t and $X_{c\infty}$ illustrates the heat of fusion respective the maximum absolute crystallinity for an entire crystalline polymer. $B(T)$ and n need to be determined empirically, preferably by DSC-analyses. To extend the model to non-isothermal crystallisation as it appears during thermoforming, the Nakamura model considers the previously already determined parameters into the following relation:

$$\frac{d(DoC)}{dt} = n\,(ln(2))^{\frac{1}{n}}\left(\frac{1}{t_{1/2}}\right)(1 - DoC)\,(-\,ln(1 - DoC))^{\frac{n-1}{n}}.$$

In this context, the crystallisation half time $t_{1/2}$ represents the time at which the extent of crystallisation is 50% for isothermal conditions. The derivation of both equations can be examined in CHI94 and NAK73.

Mechanical properties (tensile strength, Young's modulus, chemical resistance)

In matters of tensile strength, merged polymer chains parallel to each other increase the strength in their direction, which again increases the strength of the whole polymer network and thus the thermoplastic part. Hence, high degrees of crystallinity are to be pursued to achieve improved mechanical properties. Talbott et al. analysed and described the mechanical performance of neat PEEK in dependence to the Degree of Crystallinity in terms of empirical analytical equations, which showed, that strength and Young's modulus can increase by ~30%, if the DoC is increased from 20% to 40% [TAL87]. She also investigated that this relation is almost linear and that it also influences fibre-reinforced PEEK APC-2. A comprehensive overview of the microstructure of crystalline structures is given by Wang et al. who also explored the influence of the crystalline structure on the elongation at break [WAN18]. Furthermore, the denser polymer network enhances the resistance against chemicals as for instance Skydrol. Disadvantage of the improved mechanical performance is a reduction in fracture toughness, which is reduced about 22% for an increased DoC from 20% to 32%.

Density

Due to denser packing of the molecular chains, the volume of the polymer network decreases resulting in a higher density for high degrees of crystallinity. VICTREX indicates for crystalline PEEK 151G a density of 1.3 g/cm^3, while amorphous PEEK has a density of 1,26 g/cm^3 [VIC12]. This must be considered, if a detailed analysis of warpage or internal stresses is of interest.

Relative Degree of Crystallinity

For identifying the degree of crystallinity, the enthalpy of fusion ΔH_m^0 for a theoretical entire crystalline spherulite is required. For PEEK, the value for the enthalpy of fusion is 130 J/g and is given by literature. However, for PEEK AE250™-polymer no values have been published yet. Methods for determining the total enthalpy of fusion are described by Ehrenstein [EHR03].

Parameters

For identifying the melting enthalpy, the neat polymer specimens were heated above melting temperature in a Differential Scanning Calorimeter DSC Q 2000 by TA Instruments. The temperature was kept constant for five minutes and then cooling was performed at a specified cooling rate to induce a reproducible crystallisation. Subsequently, the specimens were heated again above melting temperature. A heating rate of 10 K/min was chosen which is also an appropriate value in literature [EHR03]. All measurements were repeated once with new specimens.

Cooling rate dependent crystallisation

Both polymers were investigated concerning the influence of the cooling rate on the crystallisation behaviour. To create a reproducible condition, the experiments started in molten condition above T_m. During the subsequent cooling, the thermal heat flux was measured giving information about the crystallisation characteristic. Afterwards, the specimens were heated again above melting temperature at constant heating rate of 10 K/min. The melting enthalpy of both polymers with respect to the cooling rate is shown in Figure 10.

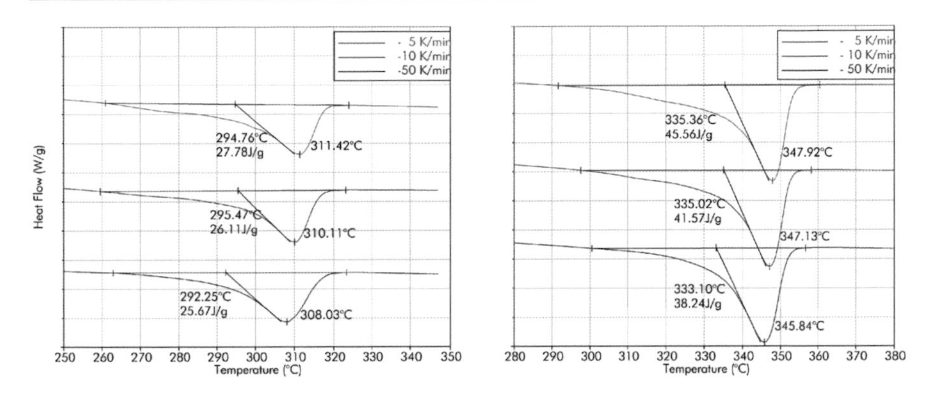

Figure 10 – Enthalpy during melting after different cooling rates: AE250™-polymer (left) and 151G (right)

The measured melting enthalpies validate the theory that increasing cooling rates reduce the creation of crystallites. It is noticeable, that the melting enthalpy of AE250™-polymer is clearly smaller than for 151G, which is reasonable due to a smaller molecular weight. Nevertheless, for AE250™-polymer the difference in melting enthalpy between a low-cooling rate of 5 K/min and a high-cooling rate of 50 K/min is relatively small with only ~2,4 J/g in average. This might result from a high degree of nucleation and crystalline growth already during cooling. The obtained results illustrate that the cooling rates exerts less influence on the crystallinity of AE250™-polymer than on classic PEEK 151G. However, its total enthalpy of melting is surely below 130 J/g. To allocate relative degrees of crystallinity, the total heat of fusion would be necessary to determine, which was not done in this context.

For 151G the obtained values of the relative DoC are significantly higher and are more sensitive to the cooling rate. Considering the enthalpy of fusion of a potentially entire crystalline polymer of 130 J/g, the degree of crystallinity of 151G is in the range between 36% and 29% for the tested cooling rates.

The measured glass transition region is almost similar for both polymers at approximately 150°C. For the melting peak, values of 309°C respective 347°C are observed in six measurements with standard deviations below 1,3 K (Table 5).

Table 5 – Averaged degree of crystallisation for different cooling rates

			ΔH_m – Enthalpy of fusion [J/g]			DoC- Degree of Crystallinity [%]		
Cooling rate [K/min]			- 5	- 10	- 50	- 5	- 10	- 50
	T_g [°C]	T_m [°C]						
AE250	151	309	28,2	27,7	25,8	22	21	20
STD [%]			6,6	2,9	2,9			
151G	148	347	47,3	43,0	37,7	36	33	29
STD [%]			3,8	3,4	3,1			

The experiments also give information about the temperature range and time-span where crystallisation takes place during cooling. Here the influence of the cooling rate can be seen in a time and temperature shift with increasing cooling rate. For both tested polymers the peak crystallisation temperatures for the small (5 K/min) and high cooling rate (50 K/min) differ for about 20 K. Since a crystalline fraction has a tremendous effect on the Young´s modulus of semi-crystalline thermoplastic polymers, the results indicate that an increasing cooling rate postpones the temperature from which the Young´s modulus of the polymer starts to increase to lower temperatures. This directly takes influence on the creation of local internal stresses. Unsymmetrical temperatures and cooling rates thus will result in asymmetries in internal stresses and therefore in warpage.

Results

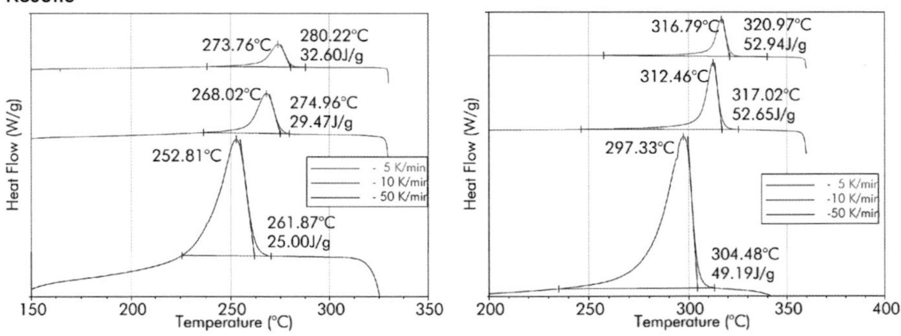

Figure 11 – Crystallisation during cooling: AE250™-polymer (left) and 151G (right)

The following Figure 12 elaborates the relationship of the cooling rate and the Peak Crystallisation Temperature, which are extracted from Figure 11. The values clearly illustrate the shift of the peak crystallisation temperature to smaller temperatures at increasing cooling rates, which corresponds to results published in literature. For increasing cooling rates, this trend will continue [EHR03]. Furthermore, for both polymers in the range from 5 to 50 K/min, the relation between cooling rate and peak crystallisation temperature is almost linear. For the developed thermoforming process, which is introduced in chapter 5, cooling rates are in the range between -190 and -20 K/min throughout the process.

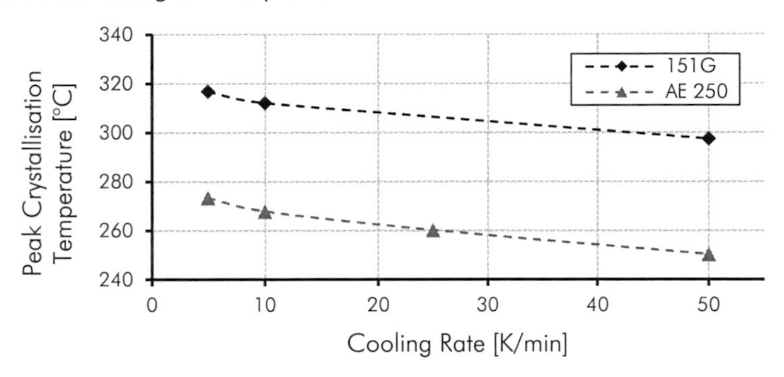

Figure 12 – Peak crystallisation temperature at different cooling rates

Process window for annealing

A promising method to find a compromise between fast cooling rates and a high degree of crystallinity is subsequent heating of the specimen at temperatures above T_g, which can support the process of post-crystallisation.

Recent experiments by Regis investigated the annealing of filled and unfilled PEEK [REG17]. They investigated an annealing temperature of 250 - 270°C which resulted in a significant increase of DoC in dimensions of 15% combined with increasing complex Young's modulus in dimensions of +6% for neat PEEK and +8% for 30% filled PEEK. In addition, the flexural strength was shown to be improved by ~9%. However, for laminates annealing temperatures above 240 °C are not recommendable, due to the resulting low Young's modulus, which provokes void formation caused by dissoluted air.

Conclusion

In order to conclude, the degree of crystallinity of both PEEK-based polymers is directly linked to the cooling rate. The manufacturing process should therefore propagate homogeneous cooling rates to prevent different solidification points or crystallisation behaviour, in order to minimise internal stresses respective warpage. For neat polymers, the cooling rate has a large influence on the enthalpy of melting, although this effect is reduced for fibre-reinforced polymers with fibres acting as impurity. The experiments further show that despite the large difference in their melting temperature, the glass transition temperature of AE250™-polymer is comparable to that of classic 151G.

3.3 Heat capacity

The specific heat capacity c_p of a material describes the amount of energy required to heat up the material about a certain temperature change at constant pressure. It is a necessary material parameter for calculating thermal processes as the thermal heat transfer for example.

Literature and models

The specific heat capacity can be affected by the current temperature or the DoC, ageing effects, co-polymers or influences from specimen preparation and measuring parameters and methods.

At low temperatures, polymers normally have a reduced specific heat capacity, since the small distances between the molecules only provide small space for motion on molecular level. The feasibility to store heat consequently increases with rising temperatures and thus increasing distances at equal weight. For the same reason, the capacity decreases with rising degree of crystallinity. However, it is assumed that temperature is the governing parameter, which is the reason that the influence of crystallinity is neglected here.

Analysis principles

Various measuring principles are feasible for determining the specific heat capacity of polymers or compounds. An adequate facility is a temperature modulated differential scanning calorimeter (TMDSC), which is described in the following. Its method for measuring the specific heat capacity is based on the separation of kinetic (e.g. crystallisation and degradation) and thermodynamic (e.g. glass transition and melting) effects. Kinetic effects are irreversible and are not affected by the modulation, and thermodynamic effects are reversible and can be used for further interpretation. A modulation of the temperature by TMDSC separates these effects. In one run, it directly delivers a value for the temperature-dependent specific heat capacity by only considering the thermodynamic dependency between temperature change and heat flux [EHR03].

During analysis using TMDSC, a low-frequency sinus wave with low temperature amplitude is applied to the temperature program. For calculating the heat capacity, the currently modulated heat flux is divided by the modulated heating rate.

Parameters

For determining the specific heat capacity of both the PEEK polymers, a TMDSC Q 2000 by TA Instrument was used. Table 6 represents the parameters chosen for the analysis:

Table 6 – Parameters for TMDSC-Measurement

Temperature range	[°C]	30 – 370
Modulation amplitude	[K]	±1,19
Modulation period	[sec]	90
Average heating rate	[K/min]	5
Specimens per polymer	[1]	4

Results

The corresponding obtained values of the specific heat capacities of both polymers are presented in Figure 13.

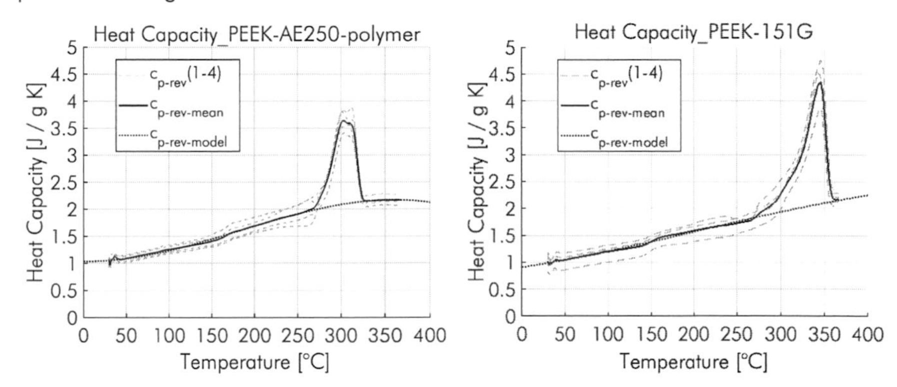

Figure 13 – Thermal heat capacity of PEEK AE250™-polymer and PEEK 151G (measured data, averaged data and simplified model)

Discussion and modelling

Both figures illustrate the increasing heat capacity with rising temperature. As stated by literature, also thermodynamic effects like glass transition are visualised adequately. A slight shift in heat capacity in the area of 150°C illustrates the region of glass transition, where molecules gain more mobility effecting an increased ability to store heat. The glass transition represents the transition from the brittle energy-elastic behaviour to entropy-elasticity for the amorphous fractions of the semi-crystalline polymer. For PEEK 151G, standard deviations of 4–8% were measured until melting, while for PEEK AE250™-polymer an uncertainty of 6–10% of the measured values are detected. Close to the melting region, a strong increase of the heat capacity can be ascertained, resulting from the required melting enthalpy. Afterwards in molten state, the heat capacity remains almost constant in the measured range.

For modelling, no appropriate mathematical description could be found, which covers all effects. A simplified model for the temperature-dependent heat capacity, which neglects the melting enthalpy, is represented by a third-grade polynomial equation (3.4):

$$c_p(T) = p_1 + p_2 \cdot T + p_3 \cdot T^2 + p_4 \cdot T^3. \tag{3.4}$$

The fitting parameters are listed in Table 7. Note that the temperature is given in degree Celsius.

Table 7 – Parameters for polynomial model of c_p

AE250™-polymer				151G			
p_1	p_2	p_3	p_4	p_1	p_2	p_3	p_4
0,906	2,448e-03	6,181e-06	-9,79e-09	1,038	-3,024e-04	2,795e-05	-5,106e-08

However, to consider the required heat for melting, the raw data will be considered for further modelling instead of the simplified model. Moreover, for the molten state above T_m, a constant thermal heat capacity is assumed, which is in the range of 2,2 J/gK for both polymers.

3.4 Thermal conductivity

The coefficient of thermal conductivity gives information about the required time for transferring heat through a material. During the process stages of heating to process temperature and cooling after impregnation, heat is conducted through the composite. Accordingly, the data of the thermal conductivity are of great importance to allow modelling the heat transfer.

Literature and models

Beside temperature, which is the major influencing parameter, also the degree of crystallinity as well as the arrangement of crystallites has an impact on the thermal conductivity of semi-crystalline thermoplastics. While previous research activities showed that the influence of the DoC is below 10% between amorphous and strongly crystalline PEEK, a high degree of orientation of the crystallites strongly influences the thermal conductivity (almost factor 3 between parallel and transverse direction) [CHO94]. For semi-crystalline polymers, especially, the phase change from solid to molten state is affected by the crystalline fractions due to different melting temperatures of crystallites. Above T_m, the thermal conductivity is governed by the counteracting effects of decreasing conductivity by decreasing density and increasing conductivity by increasing mobility of the molecules [SCO12].

Parameters

A Netzsch LFA 457 laser flash analysis (LFA) was used to evaluate the temperature dependency of the thermal conductivity λ_T, which requires knowledge about the thermal heat capacity, the precise dimensions of the specimen and its density, which is evaluated later on in section 3.7. The term a describes the thermal diffusivity in this context, which is delivered by the LFA:

$$\lambda_T = a_T(T) \cdot \rho(T) \cdot c_p(T). \tag{3.5}$$

In order to determine the thermal diffusivity by an LFA, the specimen inside a chamber is heated to the required temperature when a heat impulse by a laser beam is applied on one side of the specimen. The heat is absorbed by the specimen, which results in increasing temperature at the opponent side after short time. Time and temperature are measured by means of an infrared sensor. Subsequently, the data can be used to calculate the heat capacity, thermal diffusivity and conductivity [GLO13]. Measurements were taken each 10 K with three repetitions, while heating from room temperature until below melting temperature. Increasing the temperature by 10 K took 30 min. As specimens, injection-moulded PEEK plates were used. Their low degree of crystallinity corresponded best with the press process with fast solidification of molten PEEK. The measuring direction was in their out-of-plane direction. For AE250™-polymer, only one measuring cycle to 250°C could be realised (Table 8).

Table 8 – Parameters for measurement of thermal diffusivity

		AE250™	151G
Diameter	[mm]	25,4	
Thickness	[mm]	2	
Temperature range	[°C]	20 - 250	20 - 310
Specimens	[1]	1	2

Results and modelling

Due to the dominant influence of temperature and the significantly smaller influence of the DoC, equation (3.5) is only respecting temperature changes.

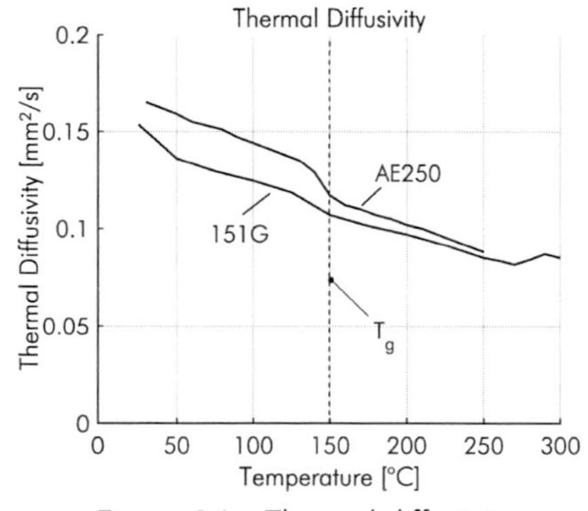

Figure 14 – Thermal diffusivity

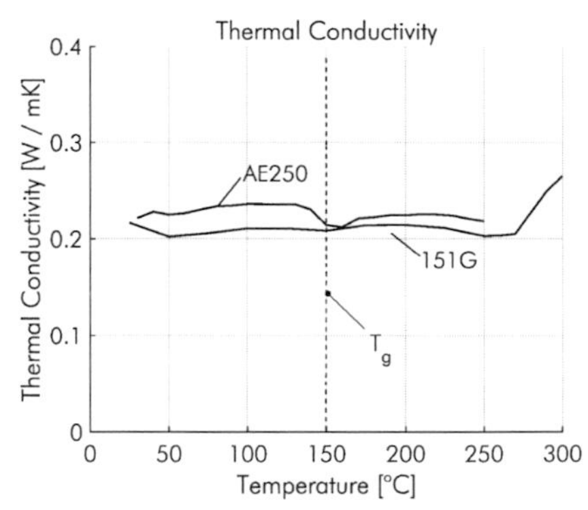

Figure 15 – Thermal conductivity

The results in Figure 14 and Figure 15 show a comparable behaviour of the thermal diffusivity respective conductivity for both polymers. While the measured thermal diffusivity decreases with rising temperature, the modelled thermal conductivity almost remains unaffected by temperature changes below melting. For temperatures above the melting temperature, the measuring could not be continued due to restrictions of the LFA. According to other technical polymers described by Dietz, a constant thermal conductivity is assumed for temperatures above T_m [DIE77]. For further modelling, a constant value for thermal conductivity of 0,21 W/mK is chosen for both polymers.

3.5 Thermal expansion

Thermal expansion and shrinking are essential properties influencing internal stresses and warpage in fibre-reinforced thermoplastic components. Especially during cooling, the creation of internal stresses is governed by the thermally shrinking matrix system, while reinforcing carbon fibres are not affected by temperature changes. Consequently, stresses occur at the fibre–matrix interface on micro-level. Especially for unsymmetrical temperature distributions during cooling, internal stresses or corresponding warpage arise. Consequently, the determination of the coefficient of thermal expansion (CTE) is of interest to examine occurring internal stresses during cooling.

Literature

The reason for thermal expansion in solid bodies is the increase of the average amplitude of atomic vibrations. This corresponds to the increase in the average value of interatomic separation, that is, the thermal expansion. For thermoplastic polymers furthermore, an increase of the CTE can be detected with rising temperature. Beside temperature, also the DoC and a preference direction of the molecule chains can take influence on the CTE. Consequently, even for unreinforced thermoplastic polymers, CTE deviations of 30% can appear for temperatures above T_g. For temperatures below, the effect decreases. Considering the DoC, already a small fraction of crystalline structures can cause an increase of CTE for temperatures above T_g [CHO90]. Below T_g, the influence of DoC on CTE is negligible.

Measuring process

In order to determine the CTE, a thermomechanical analysis unit (TMA) Q400 by TA Instruments was used. The unit consists of a heated chamber including a movable stamp which is placed with small force onto the test specimen. The specimens consisted of neat PEEK 151G. The experiments did not consider AE250™-polymer, since no injection-moulded specimen was available. For the determination of CTE, the expansion of the three specimens as detected by the stamp during heating of the chamber was considered. To provide continuous contact between the stamp and the specimen, the initial force was 1 mN, which is small enough not to counteract against the expansion of the specimen during heating. The temperature range was from room temperature to the measured crystallisation peak temperature from section 3.2.

Table 9 – TMA-parameters for measuring the thermal expansion

Material		PEEK 151G
Specimen height	[mm]	2,55; 2,35; 8,172
Force	[N]	0,001
Heating rate	[K/min]	5
Temperature range	[°C]	20 to 315

Experimental data and modelling

For calculating the CTE with respect to temperature, a central difference scheme was used, considering expansions 5 K below and above the corresponding temperature. Converting the data for elongation, the values for CTE in Figure 16 are assessed. The data verify the rise of CTE with temperature as stated in literature.

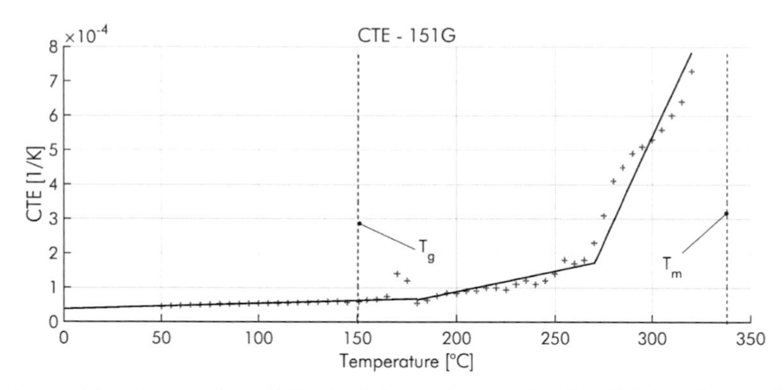

Figure 16 – Averaged coefficients of thermal expansion for 151G incl. model

Below the glass transition region, the CTE almost remains constant. At rising temperatures, the CTE increases almost linear until temperatures above 270°C. For the creation of internal stresses during cooling, the region below 270°C is of major interest. Above 270°C, a remarkable increase of CTE is detected, which has less impact on process-induced internal stresses due to the low Young's modulus of the polymer at this temperature level.

For modelling the data, an approach referring to linear equations for the regions below and above T_g and above 270°C is considered. Since it is of higher importance that the model fits the CTE for temperatures below the crystallisation peak temperature, these values are of special interest:

$$CTE(T) = a \cdot T + b. \tag{3.6}$$

Results

The corresponding model parameters are listed in Table 10.

Table 10 – Model-parameters for the CTE of 151G for different temperature stages

Temperature range	$CTE\ (T)\ [1/K]$
$T < 180°C$	$0{,}1665 \cdot 10^{-6} \cdot T + 37{,}72 \cdot 10^{-6}$
$180°C < T < 270°C$	$1{,}192 \cdot 10^{-6} \cdot T - 150 \cdot 10^{-6}$
$T > 270°C$	$12{,}288 \cdot 10^{-6} \cdot T - 3148 \cdot 10^{-6}$

Conclusion

For AE250™-polymer, comparable results can be expected. Presumably, the increase of the CTE values for the third temperature stage starts below 270°C or is little steeper. The results show that the CTE of the polymer is highly temperature dependent and that above the glass transition temperature, a drastic increase of its value must be considered. This trend increases with further rising temperature.

3.6 Flexural storage modulus

Measuring the correlation between temperature and the flexural storage modulus gives information about the resistibility against loads, which is of great importance during the phase of demoulding of a composite part. Furthermore, understanding the creation of internal stresses during the cooling step of a composite material requires knowledge of the temperature-dependent storage modulus. By correlating it with the thermal expansion, currently induced internal stresses could be assessed.

Experimental data and modelling

In order to determine the critical temperature regions during cooling of a composite, a dynamic mechanical thermal analysis (DMA) experiment was performed using a three-point bending test setup. For this, specimens were clamped into the DMA Q800 (TA Instruments) with an applied deflection of 20 μm while being heated constantly to the crystallisation peak temperature. Measuring the excitation force delivered the storage modulus, which is comparable to the tensile Young's modulus according to Jaroschek for a neat polymer [JAR12]. To investigate the influence of annealing, two specimens were heated prior testing in an oven to 220°C for three hours to provoke post-crystallisation and probably relaxation of internal stresses.

Table 11 – DMA-parameters for examination of flexural modulus of PEEK 151G

Specimen dimensions	[mm x mm x mm]	35 x 10 x 2
Drive Force stress	[N]	1,7 (corresp. 1% of bending strength)
Excitation frequency	[Hz]	1
Temperature range	[°C]	20 - 300
Heating rate	[K/min]	3
Specimens	[1]	2 + 2

The data shown in Figure 17 illustrate a remarkable change of the modulus between 140°C and 175°C, which is the glass transition region. An interesting characteristic is the increased modulus for the annealed specimens. Here, the values are increased in dimensions of approximately 300 MPa over a wide temperature range until 270°C. It is noticeable that only specimens from 151G are investigated. Specimens made of AE250™-polymer however should have a comparable characteristic.

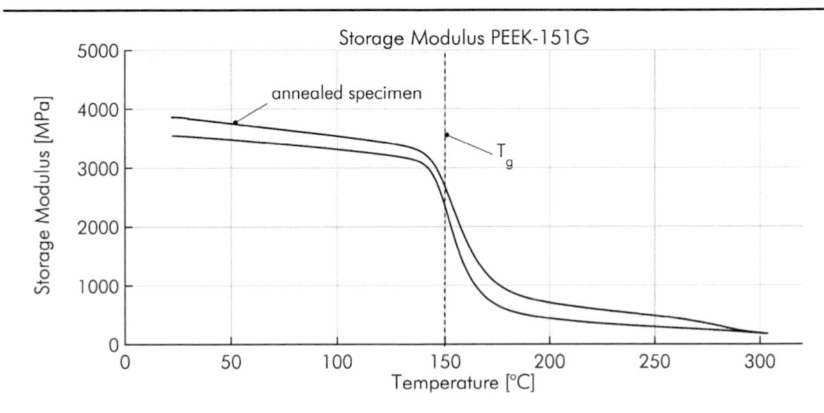

Figure 17 – Storage modulus of 151G

Conclusion

Concluding, for demoulding, the according temperatures shall thus be below 140°C to prevent deformation of the part. For reliable modelling of the process-induced internal stresses, the determined values must be linked with the DoC and CTE. This correlation is not done here, because the intention is to receive information about the temperature and time, where internal stresses occur during the phase of cooling and not to model their magnitude.

3.7 Density

Density has a direct impact on the thermal characteristic, which is the reason it is discussed below. For determining the polymers density, it is presupposed that it is not affected by the acting pressures during compression moulding. This simplification was shown to be valid by Nilsson [NIL13].

Literature

To describe density, respective the specific volume, the Multi-Tait model is suitable for the entire experimental pressure–temperature range. At acting pressures from zero to 20 MPa, the variation of the density is below 1%, which underlines the simplification of neglecting the pressure influence. The according scalar parameters a, b, c and d are provided by Nilsson [NIL13]:

$$\rho(T) = \frac{1}{a\, e^{bT}\left(1 - 0{,}0894\, ln\left(1 + \frac{p}{c\, e^{dT}}\right)\right)}. \tag{3.7}$$

However here, the densities of the PEEK polymers are modelled by inverting data of the thermal expansion from section 3.5 for PEEK 151G without considering equation (3.7).

Since no neat AE250™-polymer specimen was available, these data are only modelled based on the results of 151G, considering that the melt densities are equal and that the polymer is already molten at 305°C. The density values are consequently shifted to lower temperatures as shown in Figure 18.

Experimental data and modelling

The results point the region of breaking crystalline structures, which is clearly identified around 250°C, as already seen for the thermal expansion. In consequence, the density acts almost linear with temperature until approximately 270°C where a drop of density can be observed. For temperatures above 300°C, no data could be determined. Here, the value of 1,15 g/cm³ given by VICTREX for the fluid density is considered.

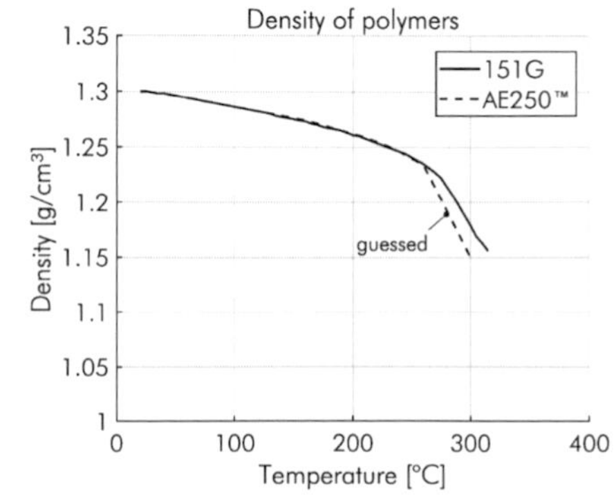

Figure 18 – Modelled density of polymers with respect to their temperature

3.8 Conclusion – PEEK polymers

This chapter characterised two PEEK-based polymers concerning their material parameters under different ambient conditions and processing parameters. In this context, the analysis parameters focussed on conditions as they exist for compression moulding processes for composite materials.

In contrast to 151G, AE250™-polymer provides a melting region which is about 40 K below that of classic PEEK, while maintaining the same glass transition temperature at 150°C. Both polymers are moreover comparable concerning their heat capacity and thermal conductivity. Concerning the crystallisation characteristic, remarkably smaller enthalpies of fusion are determined for AE250™-polymer, which might be a result of a probably lower molecular weight.

In order to promote impregnation, the reduction of the polymer viscosity is a suitable method. This chapter shows that for 400°C respective 360°C for AE250™-polymer, viscosity is in regions below 250 Pa s, which is an advantage for impregnation.

For both polymers, the cooling rate takes considerable influence on the creation of crystalline structures. While cooling rates below 10 K/min deliver comparable values for the DoC, higher cooling rates effect a decrease of DoC for both polymers. A compromise between high DoC and fast cycle times thus can be achieved for cooling rates between 10 and 50 K/min. Since no nucleation caused by fibres is considered in the conducted experiments, this is a conservative recommendation.

The investigated flexural storage modulus provides the recommendation to perform demoulding at temperatures below 140°C. The matrix modulus, and, consequently, the dimensional stability after cooling, is not given for temperatures above the glass transition region. Furthermore, an increase of the flexural storage modulus in dimensions of 300 MPa was effectuated by annealing of the thermoplastic polymer.

Heat capacity and thermal conductivity are also investigated and provide essential data for further modelling of the heating and cooling behaviour of thermoplastic fibre-reinforced laminates. For consideration of the temperature-dependent density, previously characteristics of the thermal expansion are used.

The investigated polymer data will further help to understand and to model the thermal and impregnation behaviour during consolidation of hybrid textiles to fibre-reinforced thermoplastic composites.

4 Process-dependent material behaviour – hybrid textiles

In order to capture a representative range of existing configuration of hybrid textiles, four different specifications are analysed in this chapter concerning their process-influencing parameters (Figure 19). By analysing configurations with side-by-side deposition and pre-mixed fibres, the experiments embrace the relevant available yarn structures from a low to a high degree of pre-mixing respective commingling. Comparing preforms manufactured by tailored fibre placement (TFP) to those from NCF processes also delivers information about the influence of the manufacturing method on the impregnation behaviour.

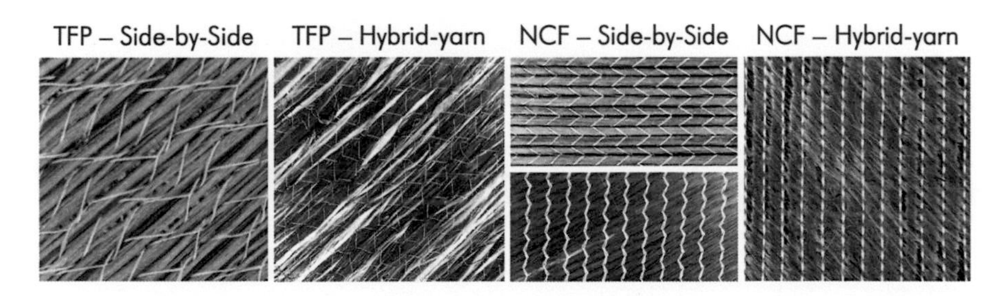

Figure 19 – Textile configurations (Pre-3 to Pre-6, f.l.t.r.)

In particular, this chapter addresses the following properties:

- acceptable shrinkage of thermoplastic matrix-yarns
- out-of-plane thermal conductivity of the textiles
- degree of mixing by distances between thermoplastic and carbon fibres
- response of the saturated carbon fibre network under compression load.

The analysed textiles are specified in Table 12.

Table 12 – Overview of analysed textile configurations

Declaration	Pre-3	Pre-4	Pre-5	Pre-6
Textile configuration	TFP-Preform	TFP-Preform	NCF	NCF
Textile-manufacturer	FIBRE	FIBRE	Karl Mayer	SAERTEX
Carbon-fibre	HTS45 P12 12k (Teijin)	Carbon 12k (unknown)	HTS45 P12 12k (Teijin)	Carbon 12k (unknown)
Thermoplastic fibre (Manufacturer)	Victrex PEEK 151G, 500 tex (FIBRE)	PEEK 600 tex (unknown)	Victrex PEEK 151G, 500 tex (FIBRE)	PEEK 600 tex (unknown)
Hybrid-yarn structure	Side-by-side	Comfil 10056 Hybrid-Yarn	Side-by-side	Hybrid-Yarn (unknown)
Global FVC	60%	49%	55%	49%
Ply-Orientation	+/- 45°	+/- 45°	+/- 60°	90°/-45°/0°/-45°
Plies per textile	2	2	2	4
Textile areal weight	386 g/m²	326 g/m²	370 g/m²	852 g/m²

4.1 Thermal shrinking

Thermoplastic yarns are manufactured in a melt-spinning process. The molten thermoplastic material is pressed through die plate perforated by small holes, so that fibres can be drawn subsequently. The drawing process provokes an alignment of molecular chains while a subsequent heat supply by calender rolls supports post-crystallisation and thus takes influence on the behaviour as matrix fibre in a compression moulding process.

Shrinking

In order to ensure a stable process throughout consolidation, the shrinking behaviour of the hybrid yarns must be considered, since thermal shrinking can lead to a displacement of the hybrid preform inside the tooling cavity during heating above T_g. Shrinking of thermoplastic yarns can be supressed by compression during the heating phase of compression moulding. However, strongly compressed yarns tend to disorientate the reinforcement fibres when melting begins. Therefore, it is essential to use shrinkage-optimised yarns to minimise the required pressure during heating. Shrinking is induced by reorientation of the semi-crystalline fractions inside the polymer during rise of their temperature above T_g on the one hand and post-crystallisation on the other hand. This phenomenon must be reduced to a minimum, which is accomplished during the spinning process. An overview of the melt-spinning process is given by Fourné [FOU95].

The absolute effect of shrinkage primarily depends on the ratio of drawing during spinning, the degree of crystallinity and the yarns structure. Here a compromise between the tensile strength of the thermoplastic fibres, supported by the drawing ratio, and the degree of shrinkage, supported by small drawing ratios, needs to be found. A reduction of shrinkage is achieved by relaxation of the semi-crystalline structure during spinning. Beside the crystalline structure, the yarns design takes immediate influence on the degree of shrinkage. In terms of the structure of a hybrid yarn, stretched yarns are more prone to shrinkage than commingled yarns, which are part wise curled and even partially broken sometimes.

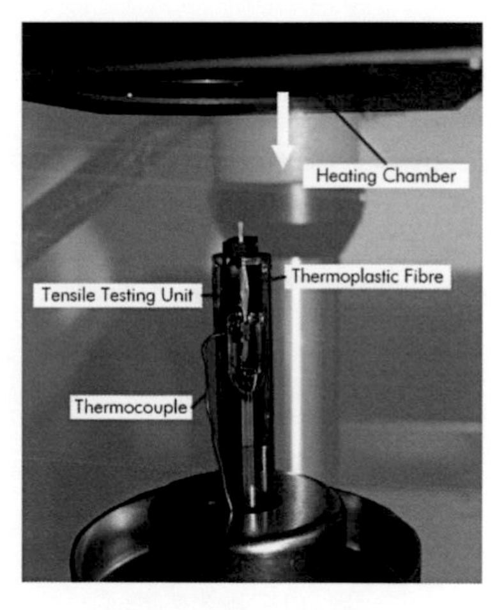

Figure 20 – TMA setup

Measuring process

A thermal mechanical analysis (TMA, Q400 TMA by TA Instruments) was used to determine the temperature-dependent thermal shrinkage. In this context, a screening of commercially available thermoplastic yarns and those spun by FIBRE was conducted to obtain an overview of feasible thermoplastic fibres for hybrid yarns and for delivering allowables for shrinkage. The yarns were clamped into a tensile testing unit inside a heated chamber. During heating, the applied tensile force was 0,025 N. The tested yarns are introduced in Table 13. Beside two commercially available hybrid yarns, also a spun fibre by FIBRE was investigated.

Table 13 – Thermal shrinkage of thermoplastic yarns

Yarn Manufacturer	Material	Specification of thermoplastic yarn [tex, filaments]	Measured elongation [%]	
			below T_g	above T_g
FIBRE	151G	385 f100	0,3	-0,75
Comfil	57C-PEEK-1400	600 f-	-0,3	0,8
Schappe	Carbon-PPS NM32	-	-0,05	-0,1

Figure 21 demonstrates the influence of the temperature and the yarn structure on the shrinking characteristic. The analysed commingled yarns by Comfil and Schappe contain thermoplastic and carbon fibres. The Schappe yarn furthermore is partially broken, resulting in a negligible dimensional change. In contrast, the thermoplastic multifilaments by FIBRE show shrinking in dimensions between 0,8% and 2%. The values for the Comfil hybrid yarn are in-between. All investigated fibre specifications provide adequate values for application as yarns for hybrid textiles, which was shown in consolidation experiments. This leads to the recommendation to allow shrinking for thermoplastic matrix yarns for hybrid textiles in magnitudes of 1%.

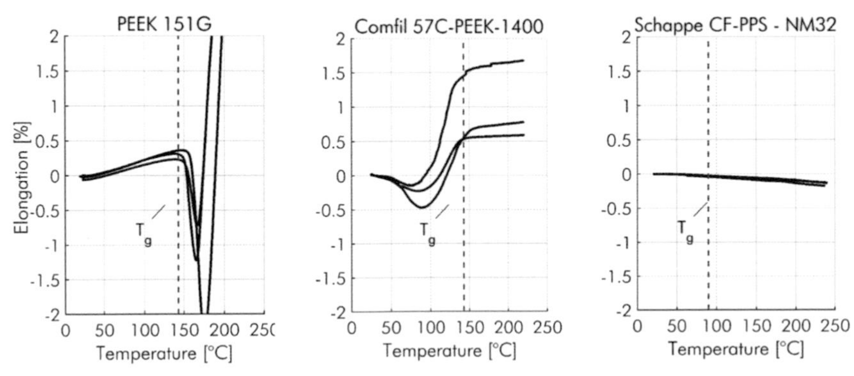

Figure 21 – Thermal shrinking of thermoplastic- and hybrid-yarns

4.2 Flow path lengths and microstructure

As stated by a simplified form of Darcy´s description of impregnation, the impregnation time increases with the square of the flow path length z as illustrated in equation (4.1). This makes it essential to determine the distance between the raw thermoplastic matrix fibres and the carbon fibres inside the preform, because different textile structures result in different flow path lengths and thus in different recommendable process parameters:

$$\Delta t = \frac{\eta \cdot z^2}{2 \cdot K \cdot \Delta p}. \qquad (4.1)$$

In order to achieve satisfying results during the impregnation process, the following prerequisites need to be fulfilled: (1) Concerning the hybrid textile, reinforcement fibres and the matrix fibres need to be stacked on top of each other at best in the direction of the acting pressure force. (2) Additionally, the thickness of the individual plies needs to be minimised. Further influencing parameters for the ply thickness are the diameter of the matrix fibres and the degree of fibre spreading of both fibre types.

Table 14 provides an overview about the microstructure of the analysed textile structures. The micrograph sections of the hybrid preforms are created by manufacturing hand lay-up laminates with thermoset resin, so that the distribution of thermoplastic and carbon fibres can be analysed. The laminates were manufactured under pressure of 0, 1, 2 and 4 MPa to investigate the pressures influence on flow path lengths. While the laminates with consolidation pressures of 2 and 4 MPa do not show any significant differences, low pressures of 0 and 1 MPa result in higher ply thicknesses in the range of 10–15%.

For analysing the microstructure of the textiles, a significant characteristic is the degree of mixing of the material. Especially, the preforms consisting of pre-mixed fibres (Pre-4 and Pre-6) benefit from the fibre arrangement, which might also be a result of the small diameter of the matrix fibres. Here, the thermoplastic filaments have diameters between 15 and 20 μm, which facilitates a high degree of pre-mixing. Nevertheless, for Pre-4 it is to note that the distribution of the materials is more chaotic and partially tends to clustering. This could lead to uneven fibre–matrix distributions after consolidation. For Pre-3 and Pre-5, thick matrix fibres were collocated in a ply-wise arrangement. Especially, the NCF shows a very regular distribution; however, the thickness of the matrix filaments results in a low degree of mixing. The TFP preform Pre-3 shows a good spreading of the matrix fibres, but suffers from partially missing matrix fibres, which presumably influences the impregnation in a negative way.

Especially in case of uneven matrix fibre distribution, it is recommendable to minimise the acting pressure during the melting stage to keep open channels for matrix distribution.

Table 14 – Micro-Sections of hybrid textiles in unimpregnated state at 4 MPa pressure (Pre-3, -4, -5, -6 from top to bottom; matrix-fibres in red; carbon-fibres in white)

	Preform ply thickness [mm]	TP-filament-diameter [µm]
Pre-3 TFP-Preform side-by-side	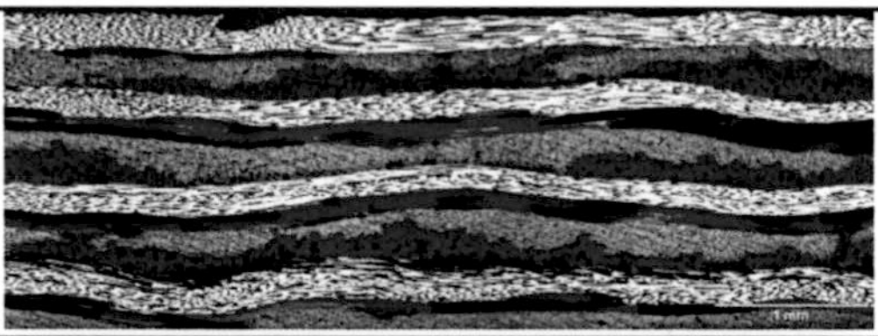 0,38 @ 0,1 MPa 0.34 @ 4 MPa	50
Pre-4 TFP-Preform pre-mixed rovings	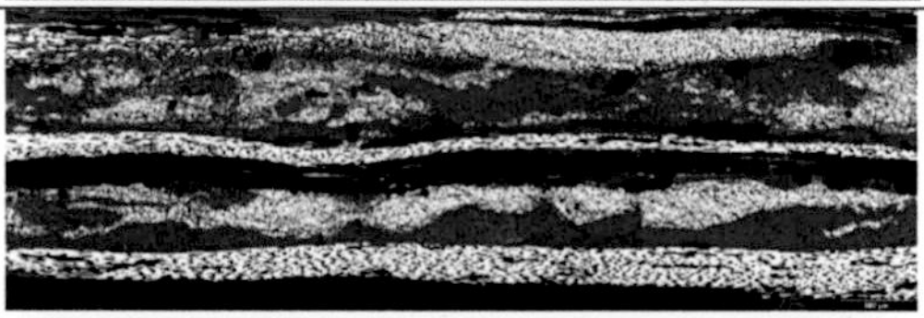 0,52 @ 0,1 MPa 0,43 @ 4 MPa	20
Pre-5 Hybrid NCF plywise integration	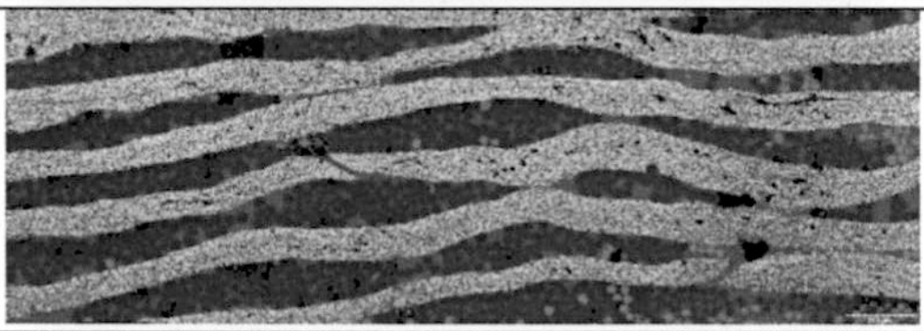 0,48 @ 0.1 MPa 0,46 @ 4 MPa	50
Pre-6 Hybrid NCF pre-mixed rovings	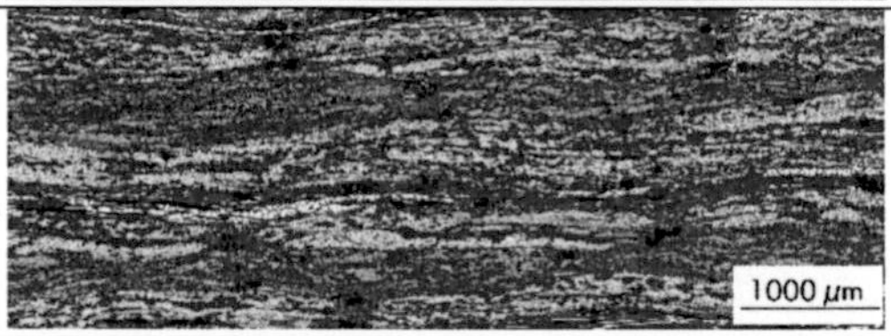 0,36 @ 0,1 MPa 0,33 @ 4 MPa	15

For modelling the impregnation characteristic, the impregnation length is an important parameter, which will be expressed in a simplified form by a scalar value in chapter 6. It represents the longest observed distance for a flow path length between thermoplastic fibres and carbon fibres and thus gives a conservative estimate.

The evaluated flow path lengths are shown in Figure 22. To investigate the values, three single plies per textile have been analysed by micro-sections at seven different points with

1 mm distance. The bright bars represent the longest mean thickness of a carbon fibre section inside a ply without being interrupted by thermoplastic filaments. The mean ply thicknesses are illustrated by the dark bars.

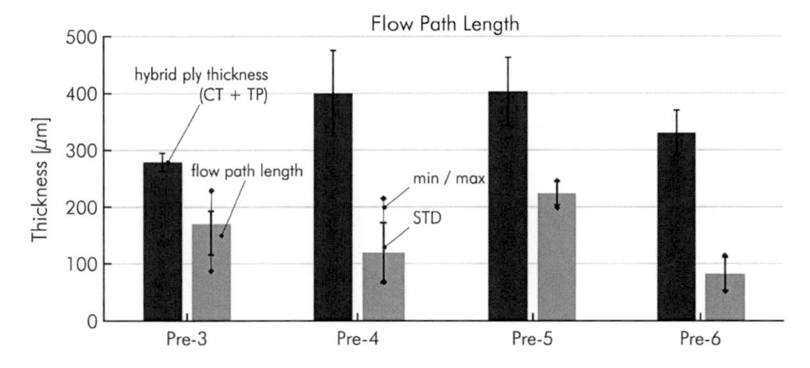

Figure 22 – Investigated ply thicknesses of hybrid textiles incl. standard deviation and min/max values

Examining these data delivers an understanding of the dependency between pre-mixing of the fibres and the resulting flow path length. While for the pre-mixed fibres in Pre-4 and Pre-6 the resulting flow path length is in dimensions of 100 μm, a ply-wise arrangement as it can be seen in Pre-3 and Pre-5 leads to larger distances to cover for the thermoplastic melt. In this context, the coarser mixed hybrid yarn in Pre-4 also has a high relative standard deviation, which presumably leads to local fluctuations of the required impregnation time. According to flow path lengths from 170 to 220 μm, Pre-3 and Pre-5 might require the longest durations for impregnation.

For modelling the impregnation in chapter 6, the values for the impregnation length act as input parameters.

4.3 Conductive heat transfer

In order to capture the complex problem of heat transfer in a hybrid textile, the carbon fibres and the thermoplastic fibres respective the thermoplastic melt need to be considered in a thermal model. For modelling the process of heating, it is necessary to distinguish between unimpregnated and impregnated state since the thermal conductivity might increase, due to a reduction of air between the filaments and a filling with thermoplastic molten polymer.

To determine the thermal conductivity λ_T for the particular materials, a laser flash unit (LFA) as already described in section 3.4 was used. In contrast to neat polymer solids, the thermal conductivity of textiles depends on further parameters as for example the fibre volume content respective included gas (air). These parameters are adjustable by applied pressure so that compressed textiles promise increased values for thermal conductivity compared to those not exposed to pressure [EHL01]. To investigage this, a sample holder for textiles allowed adjusting the applied pressure in the further described experiments.

Table 15 represents the parameters used to determine the thermal diffusivity of the textiles.

Table 15 – Parameters for investigation of thermal diffusivity of hybrid textiles

		Pre-3	Pre-4	Pre-5	Pre-6
Sample thickness:	[mm]	2,2	2,4	2,2	2,3
Sample diameter:	[mm]	10			
Compression pressure:	[MPa]	1			
Temperature range:	[°C]	50 – 300			
Repetitions per temperature:	[1]	2			
Repetitions per sample:	[1]	2			

4.3.1 Heat transfer in carbon fibre textiles

Analysing the influence of implemented thermoplastic fibres on thermal diffusivity requires knowledge about the behaviour of neat carbon fibre textiles. To consider this, an NCF material only containing carbon fibre and comparable to Pre-3 was analysed to compare its values to those of hybrid textiles. Figure 23 illustrates the measured data and shows high fluctuations in the range of ±50% of the mean value. While a single specimen provides a relatively constant thermal diffusivity, the variations between the specimens are large. A very likely explanation for this is the varying structures on meso-scale and micro-scale, resulting from the small dimensions of the specimens. Here, it is difficult to extract congruent specimens from the textile. Thus, locally differing structures, as additional or missing rovings, lead to inhomogeneity and consequently in varying thermal heat conduction.

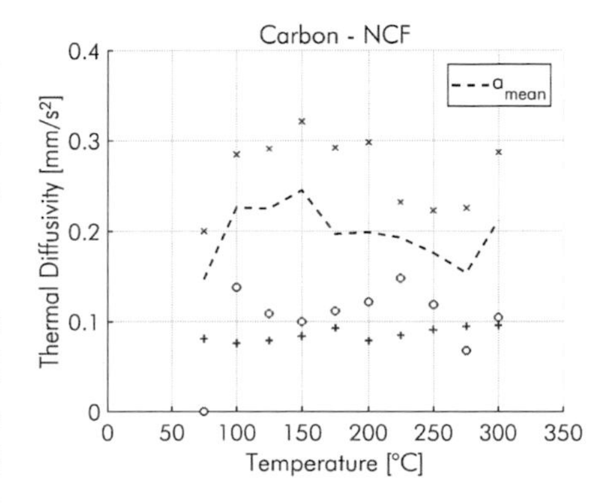

Figure 23 – Thermal diffusivity of CF - NCF

For further modelling of the thermal heat transfer inside an unimpregnated carbon fibre preform, a temperature independent thermal diffusivity of 0,2 mm/s² is considered. This value also fits the data obtained from Yang et al. or Schaefer et al. [YAN13, SCH16].

4.3.2 Heat transfer in hybrid textiles

For the dry or **unimpregnated state**, the degree of mixing can take influence on heat transfer inside hybrid textiles. Thermoplastic materials generally have a weak thermal conductivity compared to carbon fibre, which makes it interesting to quantify their effect. For preform arrangements with ply-wise deposition of thermoplastic fibres or films, the thermoplastic plies can act as isolating interface, while for preforms with a high degree of mixing, the conductive heat flow can still take place by direct contact between the reinforcement fibres. To analyse these effects, smeared values for hybrid textiles were examined in their dry state on macro-scale. For heat transfer through the dry preform, the distribution of the measured thermal diffusivity is illustrated in Figure 24.

Measuring and experimental data

Similar to the neat carbon fibre textile, the values for the thermal diffusivity fluctuate while no clear temperature dependence can be interpreted. Therefore, again scalar values are used for further modelling. An interesting observation is that the values are at an equivalent level as the carbon textile, despite the presence of thermoplastic fibres. However, Pre-3 offers a slightly better heat conduction, which probably is an effect of contact of carbon fibres. For Pre-4, Pre-5 and Pre-6 values are almost equal. Regarding the standard deviation, it points up that all values remain on a comparable level.

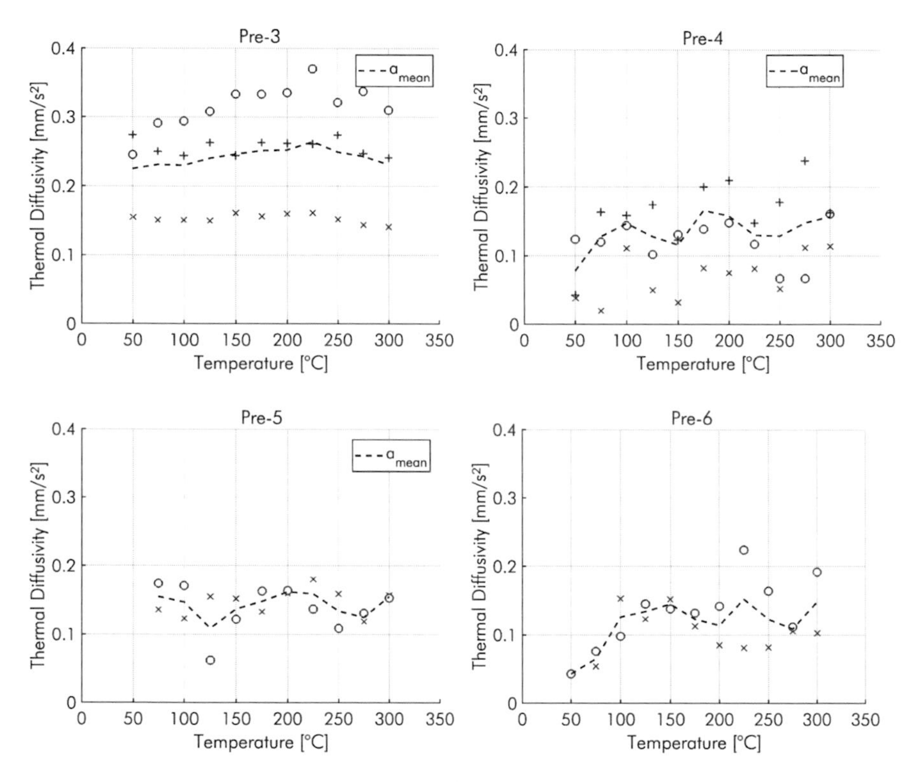

Figure 24 – Thermal diffusivity of Pre-3 to Pre-6.

Table 16 – Chosen scalars for thermal diffusivity of dry preforms

	Pre-3	Pre-4	Pre-5	Pre-6
Thermal diffusivity [mm/s²]	0.24	0.14	0.15	0.13

To address the **impregnated state**, the thermal conductivity is modelled by a combination of thermal conductivity of the carbon fibres and the molten polymer. In particular, a form following Raleigh´s expression is chosen that combines both materials for the combined value, which was already proven by Rolfes, Ehleben and Scholl to give accurate values [Rol95, Ehl01, Sch12]. It considers all material data with respect to temperature:

$$\lambda_{lam} = \frac{\lambda_{F\perp} + \lambda_M + (\lambda_{F\perp} - \lambda_M) \cdot \varphi_F}{\lambda_{F\perp} + \lambda_M - (\lambda_{F\perp} - \lambda_M) \cdot \varphi_F} \cdot \lambda_M \,. \tag{4.2}$$

For the impregnated state, the model considers observed data for the conductivities and density from sections 3.4 and 3.7 for the thermoplastic fraction. Thermal data for carbon fibres are obtained from NAS97. The modelled data, shown in Figure 25, show a direct link between the fibre volume content and the thermal conductivity. They represent the behaviour in molten state at high process temperatures.

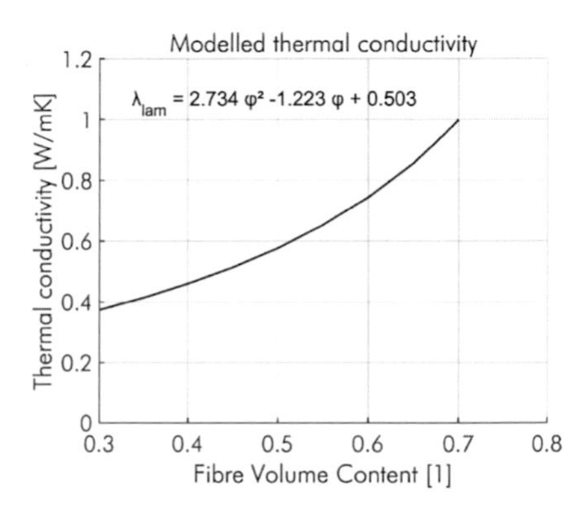

Figure 25 – Modelled thermal conductivity of laminate in molten state

Anyway, for modelling a transient heat transfer, it is necessary to obtain data for temperature and the fibre volume content. All combined parameters deliver the thermal diffusivity, as shown by equation (4.3):

$$a_{lam} = \frac{\lambda_{lam}(\varphi, T)}{\rho_{lam}(\varphi, T) \cdot c_{p,lam}(\varphi, T)}. \tag{4.3}$$

Using the experimentally obtained and modelled values from the previous chapters delivers the charts in Figure 26. For further modelling, these data are not considered in terms of equations but as direct temperature-dependent values.

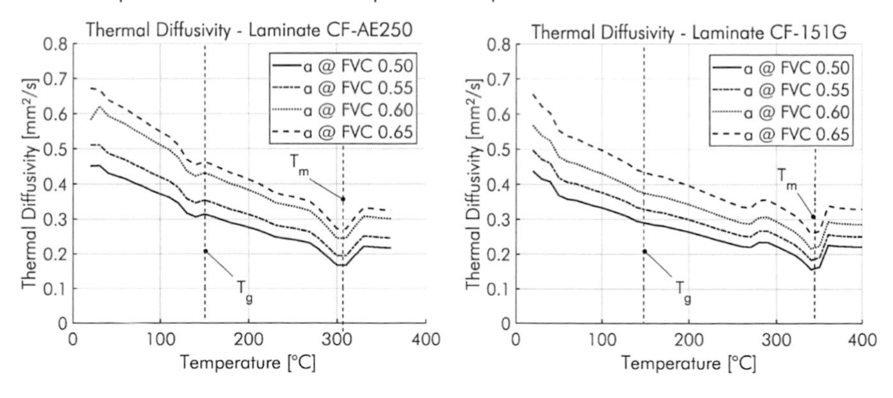

Figure 26 – Modelled thermal diffusivity of fully impregnated laminates

To conclude, thermoplastic fibres inside hybrid yarns have a small effect on thermal conductivity of hybrid textiles. Neither the textile architecture nor the temperature shows a significant influence. Impregnated composite laminates on the other hand show a clear temperature dependency. Their thermal diffusivity can be increased by up to factor 5 compared to the non-impregnated state.

4.4 Compaction characteristic

The fibre bed compaction characteristic is relevant for modelling to define the elastic packing response as a reaction on the pressure field. For a through-thickness flow, pressure gradients and viscous forces act in the same direction as the fabric compaction. This leads to a coupling between the flow and the preform compaction, resulting in an inhomogeneous distribution of the fibre volume fraction. The permeability also is directly affected by this effect, which is described in the next chapter in detail.

A special challenge for hybrid textiles is to separate the response of the carbon fibre network and the matrix fluid. Therefore, this chapter introduces an experimental method for investigating the textile network response exclusively.

Literature and models

Methods to separate the textile response from the fluid response are the testing of a dry fibre bed or the impregnation with oil after dissolving the matrix system chemically, although this can affect the arrangement of the fibres. The use of a fluid between the fibres is recommendable for the examination of the elastic response of the fibre bed since it reflects reality. During impregnation, the matrix fluid acts as a lubricant, which can reduce friction [GUT97]. Experiments by Kim or Wysocki showed that the compressive elasticity of lubricated preforms with a resin viscosity of 1 Pa s is smaller compared to dry preforms [WYS05].

Concerning modelling the preform response, different approaches exist, which mainly distinguish between linear elasticity (with or without hysteresis) and hyper-elasticity, considering a time-dependent relaxation. Several approaches are presented below.

Gutowski

The Gutowski model assumes a time invariant pressure on the fibre–matrix network and offers advantages due to its simplicity since it requires only one empirical value κ_f. It suggests to model the fibre network as an elastic medium. The description of this elastic behaviour is based on the assumption, that the fibres are slightly wavy so that they act as bending beams between multiple contact points. The net pressure experienced by the fibre network σ_F is then expressed in terms of the present fibre volume fraction φ_F, φ_{max} and φ_0, where φ_0 is the initial fibre volume fraction at which the dry fibre bed starts to carry compressive load, φ_{max} is the maximum available fibre volume fraction at quasi-infinite pressure and κ_F is the elastic constant of the fibre bed:

$$\sigma_F = \kappa_F \frac{\sqrt{\varphi_F / \varphi_0} - 1}{\left(\sqrt{\varphi_{max} / \varphi_F}\right)^4}.$$

Nonetheless, the model neglects viscoelastic effects caused by rearrangement of fibres and time-dependent saturation.

Toll

Toll and Månson conducted experiments investigating the response of different textile lay-ups as fabrics, aligned fibre networks or mats in dry and lubricated state. An effect of high compression rates in dimension of 1 mm/s was determined. For values below, the compression rates, however, showed no influence of lubrication. They proposed a power–law relation between the acting stress and the fibre volume content. Again, the empirical parameters k and n are necessary to determine [TOL95, TOL98]:

$$\sigma_F = kE(\varphi_F^n - \varphi_{F0}^n).$$

Merotte

Despite increasing complexity, Merotte et al. use a tangens-hyperbolicus-based model with one empirical, two material and one process variable. For acting pressures close to zero and for the asymptotic behaviour at high stress levels, this model delivers more accurate values than the power–law model. The same model was successfully applied by Studer et al. Apart from the others, this model delivers values for the fibre volume content with respect to the acting stresses [MER10, STU16]:

$$\varphi_F = \varphi_{F0} + (\varphi_{Fmax} - \varphi_{F0}) \cdot tanh^n\left(\frac{\sigma_F}{p_{max}}\right).$$

Wysocki et al.

Wysocki covered the topic of investigating GF-PP hybrid yarn preforms from TWINTEX raw materials. Their model distinguishes between compactions on macro- and meso-levels without coupling of these. It assumes that macroscopic tractions govern the macro-level process, while fluid pressure governs the meso-level process, so that the microstructure is only marginally affected by the preform compaction.

A pure volumetric response describes the response of the preform on macro-level, as already used by Wysocki and Toll, following the aforementioned power–law assumption. On meso-scale, the fluid pressure is the governing parameter for the fibre bundle response. Considering the fluid pressure, the elastic packing of the dry fibre bundle is described by:

$$\sigma_F = kE\left(\left(\frac{\varphi_0(e^{\varepsilon_v} - \varphi_0)}{e^{\varepsilon}(1-\varphi_0) + \varphi_0(e^{\varepsilon_v} - 1)}\right)^m - \varphi_F^m\right).$$

Here k and m act as parameters depending on the fibre arrangement within the fibre bundles. Furthermore, ε and ε_v represent the solid respective the viscous compaction strain. To isolate the fibre network response from the total response in experiments, excess resin drains into a porous insert.

Rouhi, Wysocki, Larsson

To face relaxation of the textile response, Rouhi, Wysocki and Larsson performed detailed studies. They consider hyper-elasticity by decomposing the densification during pressing into a reversible component and a non-reversible component [WYS07, ROU13]:

$$\epsilon = \epsilon_e + \epsilon_p = lg\left(\frac{\varphi_0}{\varphi}\frac{(1-\xi(1-\varphi))}{(1-\xi(1-\varphi_0))}\right) + lg(1-\xi(1-\varphi_0)).$$

Here, ϵ represents the logarithmic ratio of the density between current and fully impregnated states. The reversible component ϵ_e is essentially related to the fibre volume content φ and thus to the fibre bed compaction, while the non-reversible component ϵ_p is primarily related to the degree of saturation ξ, which is determined by Darcy´s law. A special feature of the model is the consideration of dissipation by viscous forces.

Theory of relaxation

Kim et al. give a comprehensive overview of different material specifications concerning their pressure and relaxation response in dry state. Viscoelasticity is modelled by a Maxwell–Wiechert model where fibres are represented by a spring and the lubricant by damping modules:

$$\frac{\sigma_F(t)}{p_c} = \sum_{k=1}^{n}\frac{\sigma_F(0)_k}{p_c}e^{-\frac{t\,E_z}{\eta_z}}.$$

Here $\sigma_F(0)_k$ denotes the initial stress acting on the kth element, while p_c represents the initial stress acting on the sample with η_z and E_z as viscosities and Young's modulus. This study claims that the compression rate does not influence the response of the dry textile. The results furthermore indicate that the absorption of load begins at higher fibre volume contents after a first compression. Lubrication is shown to have a reducing effect on the textile compression modulus. Comparing the loading curve to the unloading curve moreover shows a hysteresis due to relaxation of the fibre network [KIM91].

Modelling

Modelling the compression behaviour requires an equation with the applied pressure as input and the resulting bulk fibre volume content as an output. The Merotte approach delivered appropriate results. For unloading, a resulting stress on the textile acts as output, being dependent on the initial fibre volume content at compressed state, the initially applied pressure and the fibre volume content at completely relaxed state. For this, a power–law approach delivers appropriate results. Although the viscoelastic approach of Rouhi and Wysocki might be more suitable, the pressure response will be modelled in steady state. For model improvement, a hyper-elastic approach could be recommendable. The following relations are chosen for modelling:

Compression:

$$\varphi_F\,(p) = \varphi_{F0} + k \cdot tanh^n\left(\frac{p}{p_{app}}\right) \tag{4.4}$$

Unloading:

$$\sigma_F(\varphi_F\,) = p_{app} \cdot \left(\frac{\varphi_F\, - \varphi_{Fr}}{\varphi_{Fc} - \varphi_{Fr}}\right)^b. \tag{4.5}$$

In this context, φ_{F0} characterises the fibre volume content at first load absorption by the textile, k denotes the difference between φ_{Fmax} (fibre volume fraction at applied pressure p_{app}) and φ_{F0}, n is to investigate experimentally. Concerning the correlation for unloading φ_{Fr} is the fibre volume content at relaxed state of the textile after a first compression, while φ_{Fc} represents the fibre volume content at applied pressure (resulting from equation (4.4)). The constant b again needs to be determined by experiments.

Experimental investigation

In order to investigate the response of the fibre network on the applied pressure, the corresponding response of the molten matrix must be eliminated. One way to eliminate the pressure response of the thermoplastic material is to use a compression mould with a liquid-permeable cavity. To follow this approach, a compression mould was developed containing a porous male die part, which can be filled with molten matrix material so that acting forces are only absorbed by the textile. Uncertainties of this method arise through shear stresses in the liquid polymer and lubricating effects. Kim et al. found out that for low compaction rates, the fluid pressure vanishes. However, they used low-viscous oil with viscosities in ranges of 1 Pa s instead of molten thermoplastics with 300 Pa s as they appear here. To minimise the effect of viscous forces, it is recommendable to reduce the compression and opening velocity to a minimum, so that the relaxed response is given predominantly by the preform [KIM91, WYS07].

Figure 27 shows the setup of the developed compression mould. Here the 100 x 100 mm large cavity is framed by an electrically heated shear-edge tooling. Temperatures up to 400°C are feasible, enabling an entire melting of the thermoplastic fibres. To reduce pressure onto the matrix to a minimum, the stamp is perforated, so that the polymer can fill it during pressure application. To record the force and the part thickness continuously at a high resolution, the tooling is mounted inside a tensile testing machine Zwick Z250.

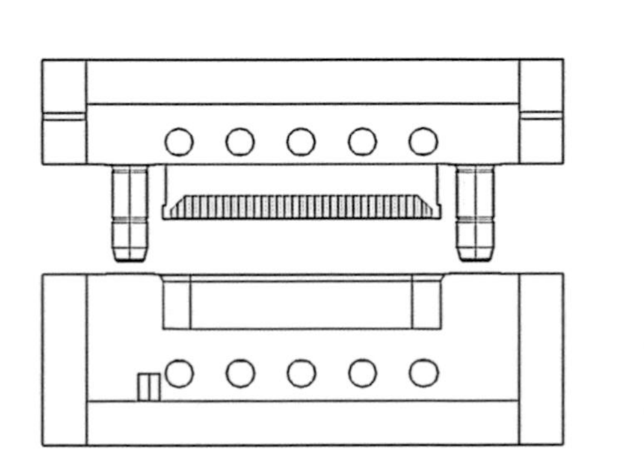

Figure 27 – Cross-section of tooling for investigating preform response

All four textile configurations are analysed concerning their compression and unloading behaviour at different acting pressures for three times per textile configuration. Each preform is loaded and unloaded four times, as shown in Figure 28. Every preform contains 16 single plies orientated in 90°-steps between the adjacent plies. An exception is Pre-6 with 45° steps between the single plies. The compression rate is chosen to 0,2 mm/min to minimise the viscous fluid fraction of the pressure response. Later derived experiments about the thickness reduction during the impregnation process, which are investigated in section 8.1, show smaller compaction rates in regions up to 0,01 mm/min. Consequently, to minimise viscous drag effects, the compression rate should be further reduced in future experiments.

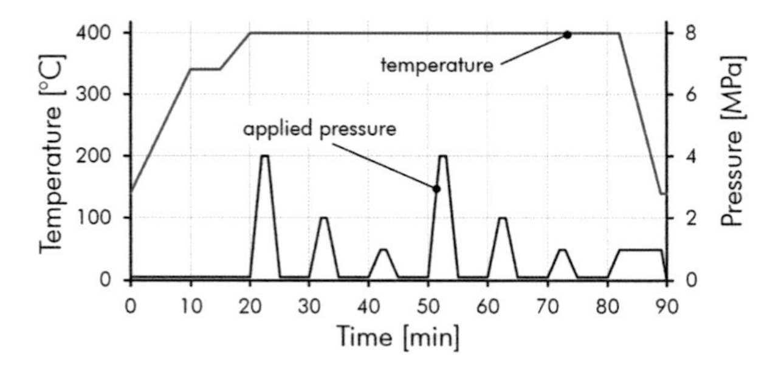

Figure 28 – Process parameters for compression and unloading experiments

Results

The following two figures illustrate the measured dependency between applied pressure and the resulting fibre volume content. In Figure 29, the compression characteristic of the textiles can be observed in a first and a second compression. During the first compression, the responses differ considerably from the subsequent iterations, as especially Pre-3, Pre-4 and Pre-5 require load already for low fibre volume fractions. A plausible explanation is that during first compression, the predominant fraction of impregnation takes place, which requires the liquid polymer to flow and thus pressure force, due to viscous forces. After the first compression, no more relevant flow occurs, so that subsequent compression–unloading cycles become reproducible. With a higher degree of mixing, flow only takes place on meso- or micro-scale where viscous forces are minor as shown by the small hysteresis in Pre-6.

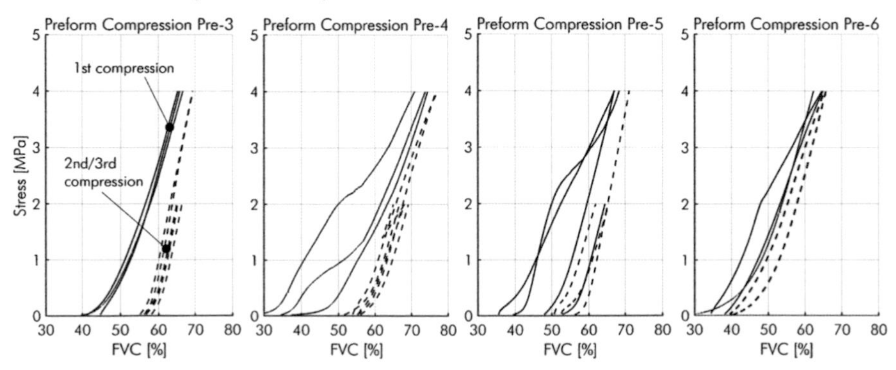

Figure 29 – Measured compression response of preforms (first compression – solid line, second / third compression – dashed line))

The results indicate that after a first compression, the response of the textile almost remains unchanged with further iterations of compression or unloading. A slight increase of the fibre volume content in the range of +3% can be detected for the first repetition of unloading.

Compression – Unloading

Comparing the compression and the unloading curves display a hysteresis between compression and unloading (Figure 30) for all textiles beside Pre-6, which consists of highly pre-mixed hybrid yarns and where the energetic loss is almost negligible. The other textiles with their lower degree of mixing face a substantial hysteresis, which presumably is a result from the process of impregnation. The results indicate a significant influence of the textile configuration on the resulting pressure response.

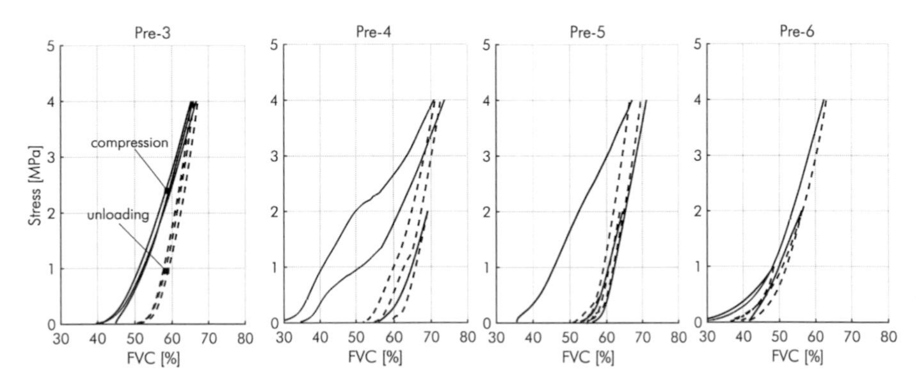

Figure 30 – Measured compression–unloading response of preforms (compression – solid line, unloading – dashed line)

The measured correlations between the fibre volume content and the stress onto the textile are processed with the previously described models, which are visualised in Figure 31.

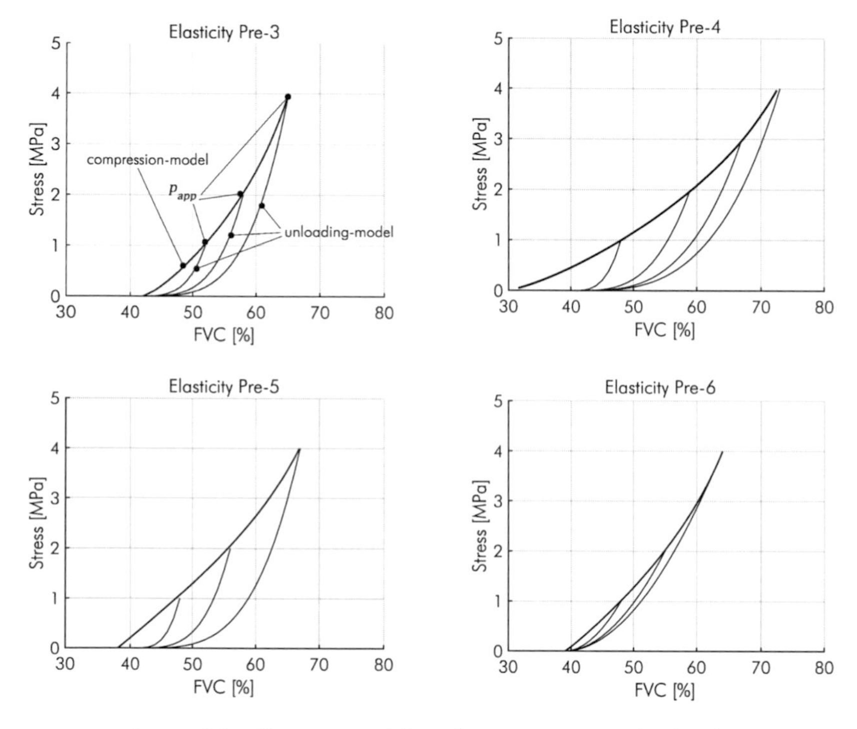

Figure 31 – Elasticity models incl. compression and unloading

The corresponding parameters for modelling are listed Table 17 and Table 18:

Table 17 – Parameters for textile compression model (eq.(4.4))

	Pre-1	Pre-3	Pre-4	Pre-5	Pre-6
φ_{F0}	25	42	38	38	39
k	49,93	28,46	33,96	37,83	32,05
n	0,2868	0,7633	0,7159	0,9922	0,9098

Table 18 – Parameters for textile unloading model (eq.(4.5))

	Pre 1	Pre 3	Pre-4	Pre-5	Pre-6
φ_{Fr}	40	40	40	40	39
b	3,25	4,128	3,37	3,86	1,93

A considerable deviation of the measured pressure response can occur by the toolings guiding pillars, potentially causing friction together with the bushings. This error is evaluated by measuring the force for mould closing and opening between the experiments. Summarising, the forces for guiding were below 400 N resulting in a maximum error of 4% respective 1% for applied pressures between 1 and 4 MPa. The standard deviation was evaluated, taking the compression responses into account. For the different textile configurations, the standard deviation showed the largest values at the stage of the first compression. Figure 32 illustrates the range of the standard deviations (grey) together with the modelled line (black). As already indicated for the flow path length and the thermal conductivity, for the inhomogeneously pre-mixed Pre-4, the deviations become remarkable, so that the developed model can only give rough estimates of the $p-\varphi$ relationship. Despite a regular arrangement, Pre-5 causes fluctuations, which might be a result from the long flow path lengths and a corresponding inhomogeneity during compression respective impregnation. In contrast, Pre-3 and the more regular textiles and Pre-6 show an improved reproducibility.

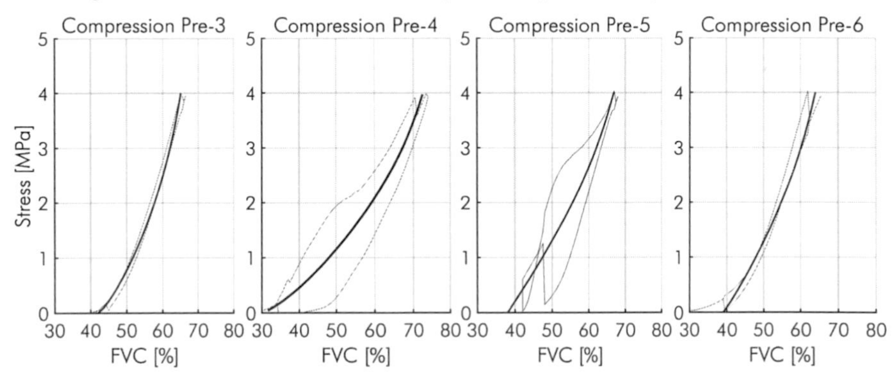

Figure 32 – Standard deviations of compression models (grey line)

Conclusion

Concluding, this chapter delivered models to calculate the compression and as well the unloading trajectory of hybrid textiles under applied pressure in molten state. The influence of the liquid matrix is clearly visible during the first compression. Subsequently, the response is comparable to that of dry textiles. Furthermore, a hysteresis between compaction and unloading almost disappears after a first compression–unloading cycle. The unloading curves, which will be more important for the developed impregnation model in section 6.2, demonstrate a high degree of reproducibility. A further result of the experiments is that for fibre volume contents in ranges of 60%, the applied pressures must exceed 2,5 MPa in three of four cases. The generated data will further be used for modelling of the impregnation characteristic in chapter 6.

4.5 Permeability

Permeability characterises the reciprocal resistance of a porous medium against a penetrating fluid. It depends on geometrical and material parameters, which usually are smeared and simplified in one value. The permeability of a porous medium depends on the fibre distribution and their particle diameter as well as the fibre volume content. Further, the surface sizing of the fibres can influence the permeability.

Literature and models

Since all influencing factors are very difficult to consider accurately, an experimental investigation of permeability is very complex. Extensive work was done in this context in a benchmark carried out by 32 laboratories across the world [MAY18].

As permeability is affected by the fibre arrangement, it is a direction-dependent property:

$$\boldsymbol{K} = \begin{pmatrix} K_{11} & K_{12} & K_{13} \\ K_{21} & K_{22} & K_{23} \\ K_{31} & K_{32} & K_{33} \end{pmatrix}.$$

Since impregnation of hybrid textiles dominantly takes place in out-of-plane direction, K_{33} is the most relevant value, which will be the only factor considered subsequently and which is further expressed by K. Beside the anisotropic behaviour, also different scales of permeability must be considered. Penetration into textile preforms contains flow around fibre bundles as well as flow into fibre bundles. Consequently, one must account for macro-permeability and meso-permeability.

Several authors that considered the lubrication theory or flow through capillary tubes developed models to predict the permeability of a textile. The analytical models by Carman–Kozeny, Gutowski, Bruschke–Advani and Gebart are reviewed briefly below, including mathematical expressions of the models in Table 19 and Figure 33.

The variation of permeability with respect to the fibre volume fraction can be predicted by the Carman–Kozeny equation (eq. (4.6)). This equation considers porosity and the hydraulic diameter in case of packed and aligned cylinders of radius r_f. All further material-dependent parameters must be determined experimentally and are expressed in terms of a single scalar value ς, which of course is a significant simplification. This is reflected in the demand for different values of ς at different fibre volume contents [KOZ27, CAR37].

Gutowski et al. improved the Carman–Kozeny model, by considering the fibre volume content in the mathematical description [GUT87]. The implementation led to a drastic decrease of the modelled permeability at high fibre volume fractions. As for the Carman–Kozeny model, there still is a need for the experimental investigation of a permeability constant k_z.

Another model to describe the permeability is based on lubrication theory and was proposed by Gebart [GEB91]. Their formulation for permeability provides the advantage that it does not involve an empirically investigated permeability constant. Instead, the permeability constant is only based on the assumed fibre arrangement (hexagonal or quadratic). Bernet et al. therefore suggest to use the Gebart model to estimate the permeability of commingled yarns [BER99]. A further advantage of the Gebart-model is the distinction between in- and out-of-plane flow in two different equations.

A further approach to model permeability was chosen by Bruschke and Advani [ADV11]. They used lubrication theory and a cell model concept to model permeability. As for the Gebart model, the Bruschke–Advani model is only influenced by the fibre radius and the fibre volume content. Generally, it is to say that modelling the process of impregnation always considers tremendous uncertainties, if a simplified permeability model is chosen.

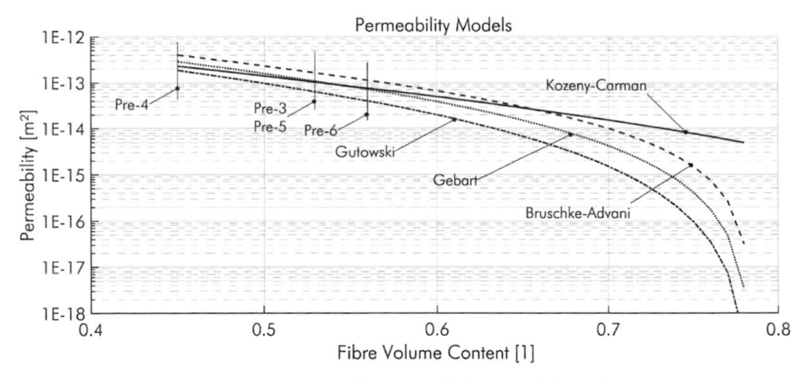

Figure 33 – Comparison of permeability models with r_f = 3.5 µm

Table 19 – Permeability models

Model	Equation	No.	Ref.
Carman–Kozeny	$K_{CK} = \frac{r_f^2 (1-\varphi_F)^3}{4 \varsigma \varphi_F^2}$	(4.6)	[Koz27, Car37]
Gutowski	$K_{Gut} = \frac{r_f^2 \cdot \left(\sqrt{\frac{\varphi_{max}}{\varphi_F}} - 1\right)^3}{4 \cdot k_z \cdot \left(\frac{\varphi_{max}}{\varphi_F} + 1\right)}$	(4.7)	[Gut87]
Gebart	$K_{Geb,\parallel} = \frac{8 r_f^2}{c} \frac{\left(1-\varphi_F\right)^3}{\varphi_F^2}$	(4.8)	[Geb91]
	$K_{Geb,\perp} = c_1 \left(\sqrt{\frac{\varphi_{F,max}}{\varphi_F}} - 1\right)^{\frac{5}{2}} r_f^2$	(4.9))	
	quadratic arrangement: $c = 57$, $\varphi_{F,max} = \frac{\pi}{4}$, $c_1 = \frac{16}{9\pi\sqrt{2}}$ *hexagonal arrangement:* $c = 53$, $\varphi_{F,max} = \frac{\pi}{2\sqrt{3}}$, $c_1 = \frac{16}{9\pi\sqrt{6}}$		
Bruschke-Advani	$K_{BA} = \frac{r_f^2}{3} \frac{(1-L^2)^2}{L^3} \left(\frac{3L \tan^{-1}\sqrt{\frac{1+L}{1-L}}}{\sqrt{1-L^2}} + \frac{L^2}{2} + 1 \right)^{-1}$ $L^2 = \frac{4\varphi_F}{\pi}$	(4.10)	[Adv11]

Measuring process

In order to compare the modelled values for permeability with experimental data, one textile specification was tested in a permeability test setup at the Institute of Polymer Engineering at University of Applied Sciences and Arts - Northwestern Switzerland (FHNW). For permeability measurement, preforms only containing reinforcement fibres are required. To prevent an influence of the thermoplastic fibres inside the hybrid preforms, separate preforms were manufactured by tailored fibre placement, without PEEK fibres and with equal specifications as for Pre-3. Thus, evaluated values for permeability should be adaptable to Pre-3. Uncertainties appear, caused by the necessity of a carrier film to stitch on it during the TFP process, which consisted of PVA. After stitching, the soluble film was washed out by water. However, it cannot be assured that no PVA residues remain inside the preforms, which could influence the permeability. Furthermore, washing also could have caused shifting of single rovings.

The permeability measurement was realised by oil flowing through the fibre network in out-of-plane direction. The pressure drop between inlet and outlet of the oil is taken as a representative value for the permeability. At the setup at FHNW, the saturated as well as the unsaturated permeability is possible to evaluate (Figure 34). Since the specimen consisted of several plies, the measured permeability must be interpreted as macro-scale permeability, which should be higher than on meso- and micro-scales. For the experiments, one preform was tested at fibre volume content of 41% and three different oil pressures.

Experimental data and modelling

Figure 34 – Setup of permeability investigation at FHNW (Source: FHNW, J. Studer)

Conclusion and error analysis

The experiments showed a reproducible measurement of permeability, as demonstrated in Figure 35. For the three fluid pressure levels at 0,05, 0,08 and 0,12 MPa, almost equivalent values were investigated close to 4,55e-13 m², which is in excellent correlation to the modelled micro-permeabilities of the Carman–Kozeny model and the Gebart model for fibre volume contents of 41%.

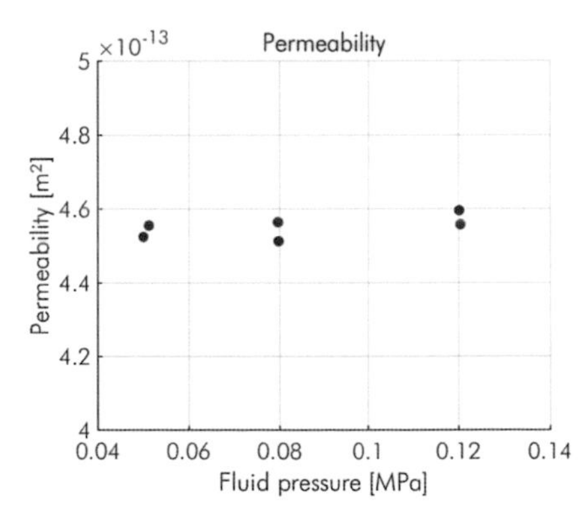

Figure 35 – Experimentally investigated permeability at FVC 41%.

The validity of the modelled values on macro-level for permeability on meso-level, as it is required for the impregnation model in section 6.2, is difficult to assess. However, the experimentally obtained data support the confidence in the described permeability models.

4.6 Conclusion – hybrid textiles

In this chapter, hybrid textiles have been analysed concerning their process behaviour during heating and impregnation. As raw materials, hybrid yarns and thermoplastic multi-filaments show to be applicable, if their thermally induced shrinkage is in dimensions below 1%.

Against the background of Darcy's equation, the flow path length is a key characteristic of hybrid textiles influencing the process conditions remarkably. Here, the degree of pre-mixing and the deposition on top of each other of thermoplastic and reinforcement fibres are essential characteristics for the impregnation process. In this context, a pre-mixed hybrid yarn provided a 63% smaller flow path length than for a side-by-side arrangement, which theoretically would lead to 87% shorter impregnation times according to a simplified Darcy equation. To improve the regularity of the material distribution in a hybrid yarn, a small diameter of the thermoplastic filaments is recommendable. Values in dimensions 20 μm or below showed to be effective.

Regarding the thermal conductivity of hybrid textiles, large variations in dimensions of more than 50% have been observed, which might be a result from varying fibre arrangements. An influence of temperature could not be investigated, in contrast to the neat polymer. For further modelling, scalar mean values are considered for the textiles.

In order to answer the question about the relationship between the textile structure, the applied pressure and the resulting fibre volume content, compression experiments were performed. It is shown that high fibre volume contents of 60% require pressures up to 4 MPa. Furthermore, the experiments illustrate that during a first compression of a textile, the viscous drag of the polymer fluid counteracts the compaction resulting in a flat FVC–strain ratio. This allows the conclusion that for a first compression of a hybrid textile at impregnation temperature, the viscous drag is not negligible, so the applied load is taken by both flowing polymer and textile. For following loading–unloading cycles, the hysteresis cycle becomes almost negligible. Furthermore, a regular arrangement of fibres and polymer leads to small deviations in the compression curve and a small hysteresis during first compression and unloading cycle. Consequently, the homogeneity of the semi-finished textile takes immediate influence on the impregnation reproducibility and thus the process stability.

According to the presented models for the textile permeability, the calculated values are highly dependent on the local fibre volume content. The Gebart and the Carman–Kozeny approaches show the best results compared to values evaluated from one experiment.

Anyway, it must still be considered that the reduction of permeability to one scalar value is a clear simplification which results in uncertainties for a further impregnation model.

Concluding, recommendations for the textile and the process design are

- to reduce shrinkage of thermoplastic fibres below 1% by small orientation of crystalline structures during spinning or to perform textile measures, for instance, a further process step to stretch broken yarns
- to minimise flow distances by
    - TP filament diameters in dimensions below 20 μm
    - a homogeneously fibre distribution, where thermoplastic and reinforcement fibres are deposited on top of each other
- to find a compromise between a high fibre volume content and a decreasing permeability, both resulting from high applied pressures.

5 Development of a two-phase thermoforming process

The use of hybrid textiles primarily focusses on maximum lot sizes up to 20.000 parts p.a. per tooling. Therefore, the addressed types of components are primarily high-performance automotive parts and even more structural aircraft components. Apart from high-reactive thermoset plastics, high-performance thermoplastic materials are the only polymer class that meets the requirements regarding processability with short cycle times and mechanical performance.

The objective of this chapter is to create an innovative alternative to the consolidation processes and their corresponding tooling techniques mentioned in the state of the art. It is obvious that no all-embracing consolidation process can be created, which meets all mentioned requirements. However, referring to Figure 7 for high-performance thermoplastic composites, there is especially still a demand for a shape-adaptive, cost-effective, safe and fast manufacturing process for complex shaped parts.

In order to achieve competitive process durations, it is a necessity to introduce and withdraw heat in short time, which is applicable for thermoplastic materials in contrast to thermoset systems as RTM6, for instance. Challenges that have to be faced for processing high-performance thermoplastics as PEEK are mainly high process temperatures up to 400°C. A conservative approach for variothermal heating and cooling is the use of oil as heating and cooling fluid. Those high temperatures, however, presuppose the use of hazardous oil, which should be avoided. Furthermore, the high temperature range can lead to considerable different thermal expansion between part and tooling, which is a problem that has to be solved.

Another objective is to exploit the potential of hybrid textiles. Here, the lightweight aspect can be improved significantly compared to classic pre-impregnated laminates (organo sheets) in terms of a fibre orientation enabling the direct transfer of loads. For the consolidation process, this means that local features, as thickness variations or curvatures or joggles, need to be realisable. Beside plane shapes, also framework structures are to be considered.

To summarise the objectives, the following key requirements are to be achieved:

- minimum masses of toolings to be heated and cooled
- process temperatures up to 400°C
- possibility of complex arrangement of fibre orientation
- 2,5D shapes of part geometry
- low costs for new cavities
- fast and reliable integration into press
- avoid thermal oil for heating or cooling of the compression mould.

5.1 Concept development

Thermoforming of hybrid textiles into near net-shape contours requires new approaches for tooling concepts, since existing solutions like for classic thermoforming or for RTM components do not maintain all positive aspects of this material class or do not fulfil the objectives. Hybrid textiles enable high heating and cooling rates, which is the reason the heating concept shall be thermally agile. Furthermore, a shear-edge concept needs to be realised, to enable compression during impregnation. These premises are framed with the need for a near net-shape part. The tasks of the process are illustrated in Figure 36.

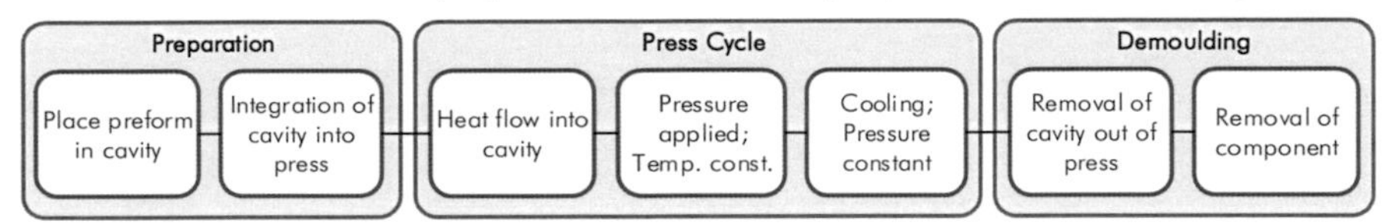

Figure 36 – Tasks of thermoforming process

The developed process principle is schematically illustrated in Figure 37, including a process diagram which illustrates the benefit in process time. Hereinafter, the process is called "Isoforming-process". Similar concepts also have been conducted successfully by the Netherlands Aerospace Centre (NLR) for two-dimensional profile geometries from chopped fibres [STE12].

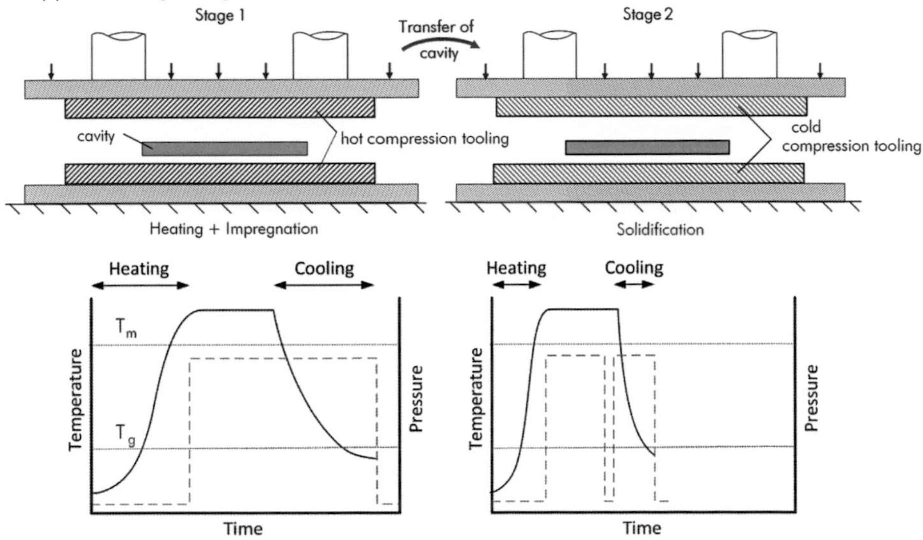

Figure 37 – Top: Two-stage concept for Isoforming process cycle; Bottom: process parameters for classic variothermal (left) and Isoforming process (right)

Thermal concept

Heating and cooling of the Isoforming thermoforming process relies on heat conduction between the cavity and two static and isothermal compression toolings. For the phase of heating and impregnation, the cavity is placed and compressed between the hot tooling, which is heated to process temperature, for example, 400°C for PEEK. At contact the temperature inside the cavity increases caused by heat conduction, while temperatures inside the compression toolings do not drop significantly due to their significantly higher total heat capacity compared to the cavity. After impregnation, the press is opened and the cavity is transferred in between the cold compression tooling. Therefore, the heat flows from the cavity into the cold tooling during compressing, so that solidification can take place.

Electric cartridges heat both the hot and the cold tooling. Once they are at process temperature, the only additionally required energy is caused by heating the cavity and from losses by natural convection.

Cavity concept

In contrast to the thermoforming processes described in the state of the art, it is characteristic for the Isoforming process that the cavity itself does not include temperature management. The shape of the cavity is kept simple and flat as can be seen in Figure 38. The intention is to reduce the tooling components that need to be heated and cooled to achieve high heating and cooling rates and to minimise energy consumption parallel.

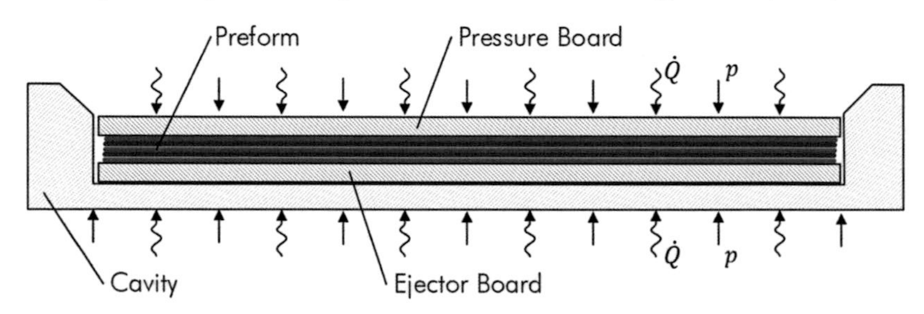

Figure 38 – Section view of cavity concept

The concept for compression relies on a shear-edge construction allowing changes in thickness during impregnation. To consider ejection of the composite after consolidation, an ejector board is integrated covering the whole bottom surface of the part. The opponent surface is also covered with a board, which transfers the load applied by the press and which prevents contact from the molten thermoplastic material to the press.

Transfer concept

For realising the transfer of the cavity between the hot and the cold toolings, three concepts for transfer have been developed as shown in Figure 39, while two of them have been realised and tested (marked with "*").

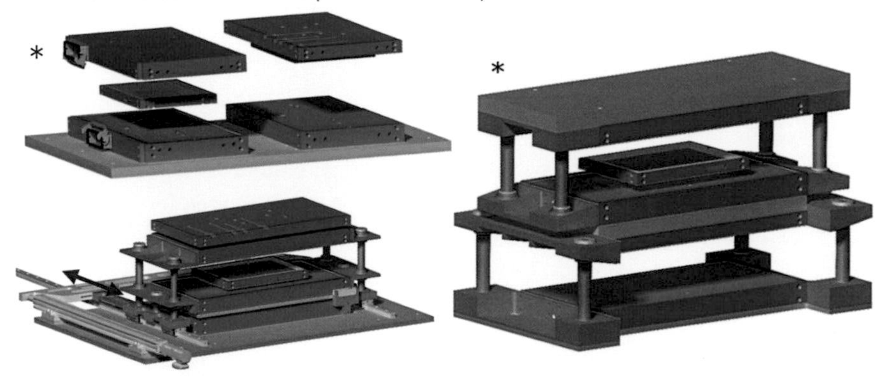

Figure 39 – Three concepts for Isoforming tooling (side-by-side arrangement; moving hot tooling; stacked arrangement) (orange = hot tooling; blue = cold tooling)

The first concept involves the isothermal compression toolings being placed side by side on the press table. For changing the stage of impregnation to solidification, the press opens and the cavity is transferred to the cold tooling. Due to the low weight of the cavity, this can be done manually or by a standard robot. However, manufacturing trials showed that this concept is only applicable with a controlled horizontal position of the press. Otherwise, an eccentric load application can cause variations of the ply thickness of the part.

In order to compete this drawback, a second concept was developed with the hot tooling being placed in-between the fixed cold toolings. For impregnation, the press closes and heating is initiated. For solidification, the press opens, the hot tooling is moved aside the press and the cavity is again compressed, this time only by the cold tooling.

A third alternative is to place the heating and cooling tools on another in a stacked arrangement. Thereby, again a central application of the press load can be realised. Furthermore, the concept allows a processing of two cavities in parallel, one in the upper and the other in the lower compression tooling. The toolings are decoupled thermally and they are aligned to each other from their centre positions. Opening is realised by springs, and the alignment between the upper and the lower toolings is due to guiding pillars. Due to its advantages, this concept was realised and will thus be explained more in detail.

5.2 Detailed design

The following paragraphs cover the detailed design of the Isoforming tooling and process considering the cavity, the design of the compression toolings and the transfer process.

5.2.1 Cavity

The cavity is made of a 1.2343 steel with good resistance to abrupt temperature changes. For the pressure and the ejector board, the same alloy is used. Steel is the beneficial material for process temperatures up to 400°C due to its comparatively low coefficient of thermal expansion which is approximately $12 \cdot 10^{-6}$ K^{-1}. However, for first trials with PA6-based composites, where maximum temperatures of 290°C were achieved, also AW7075 aluminium showed suitable behaviour.

A critical component of the cavity is the shear-edge, which allows squeezing out excessive air from the preform into the environment and which prevents molten thermoplastic material from draining out. For the two-stage thermoforming process, the shear-edge slot is between the pressure board and the cavity, as illustrated in Figure 40. The stamp of the upper compression tooling is designed smaller in dimensions of millimetres, so contact to the cavity is prevented under production conditions. Generally, a compromise needs to be found regarding a minimum gap between the pressure board and the cavity and a sufficient large gap to prevent damage to the cavity or the board during ejection or during impregnation caused by friction.

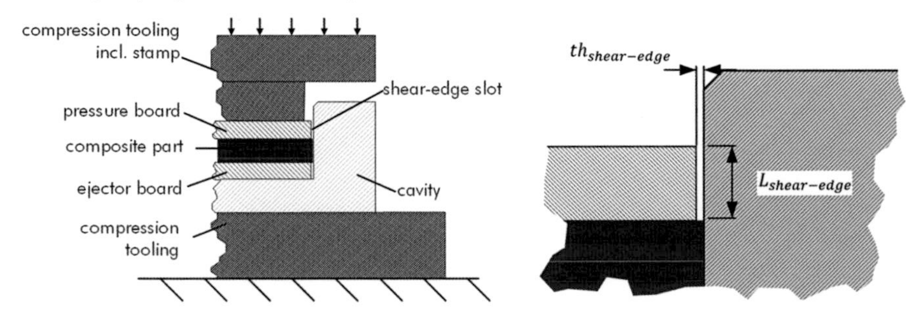

Figure 40 – Cross-section of shear-edge

With the intention of designing the shear-edge, it is necessary to consider matrix flow through the shear-edge slot during consolidation, provoked by the huge pressure gradient between the ambient and the applied pressure. For evaluating the significance of this problem, a prior prediction of the resulting flow is helpful. For this, the problem can be reduced to a flow through a thin cavity. The flow rate results from the velocity profile, which can be determined by solving the problem of a Poiseuille flow, a simplified form of the Navier–Stokes equation for stationary flow, assuming a no-slip boundary condition at the mould walls:

$$0 = -\frac{\partial p}{\partial x} + \eta \frac{\partial^2 u}{\partial z^2}. \tag{5.1}$$

Integrating and considering the boundary condition yields:

$$\frac{\dot{V}}{w_{shear-edge}} = \frac{(p_{app} - p_{ambient})}{2\,\eta\, L_{shear-edge}} \frac{4\, th_{shear-edge}^3}{3}. \tag{5.2}$$

Here $w_{shear-edge}$ describes the circumference of the shear-edge, $L_{shear-edge}$ is the depth and $th_{shear-edge}$ represents the thickness respective the distance between the tooling-stamp and the lower cavity at the shear-edge (see Figure 40). A conservative value for the viscosity of 200 Pa s is chosen with respect to chapter 3 and the corresponding values for PEEK at 400°C at shear rates below 10 s^{-1}.

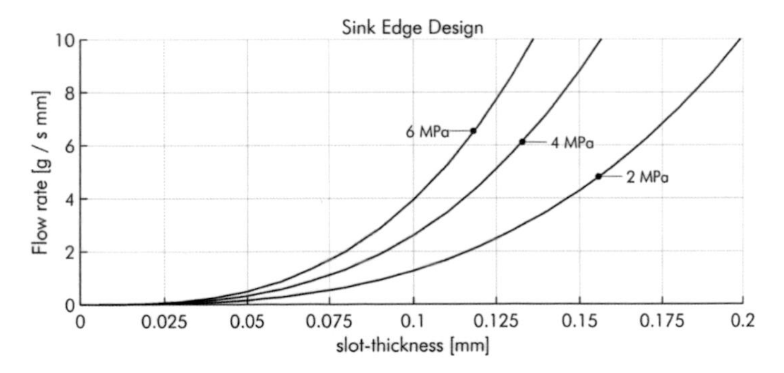

Figure 41 – Calculated flow rate through sink-edge slot

Despite the prerequisite for the Poiseuille flow of two infinitely long parallel plates, which is not the case here, Figure 41 and equation (5.2) underline the importance of minimising the slot of the shear-edge, which affects the flow rate with a power of three. In this context, a slot size of 0,1 mm is chosen giving a suitable compromise. Anyway, the pressure board is placed into the cavity without any guidance, so in a worst-case scenario the shear-edge slot can be 0 mm at one side and 0,2 mm on the other side. The same dimensions are considered for the ejector board.

For final ejection of the composite part, the cavity is transferred into a small hydraulic press where ejector pins push out the ejector board together with the composite and the pressure board. Subsequently, both surfaces are uncovered from the boards, which can be done manually.

5.2.2 Compression toolings

In contrast to classic variothermal compression toolings, there is no need for active cooling. Consequently, an isothermal temperature of the toolings can be achieved by electric cartridges. As mentioned, both the hot and the cold compression toolings are heated by electrical cartridges, delivering a heating power of maximum 2 x 21,8 kW. The arrangement of the cartridges effects a homogeneous temperature distribution on the surfaces of the bottom tooling and the stamp. All components are isolated laterally by a temperature-resistant insulation. The elements for guiding are isolated thermally, which means that their temperatures remain below 150°C.

To ascertain a reliable position of the cavity on the compression toolings, a pin is fixed in the centre of the tooling. This enables a reproducible placement of the cavity, which is equipped with a bushing in its centre at the bottom surface. By taking the centre of the cavity as reference position, it is assured that the corresponding expansion during heating or cooling does not provoke an uneven change of the shear-edge slot. On account of the large slot between the stamp and the cavity, no engineering fit links the upper hot compression tooling to the lower (Figure 42), which simplifies the construction significantly.

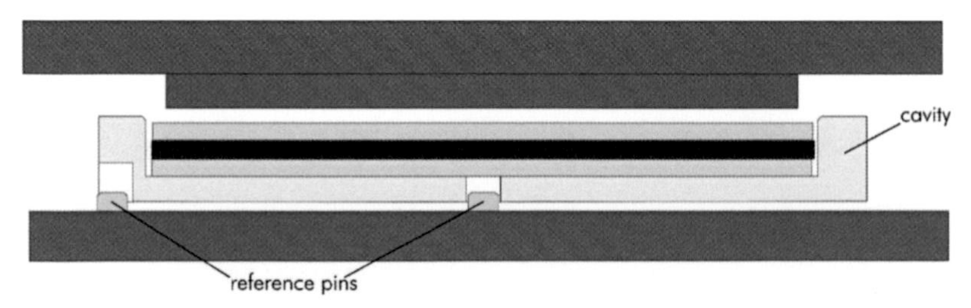

Figure 42 – Reference position for cavity

In case of new shapes to consolidate, the construction allows to disassemble the stamp and attach a new one. Since tolerances here are in the range of millimetres, the alignment of a new stamp geometry is feasible without great efforts.

5.2.3 Transfer of cavity

Depending on the dimensions of the cavity, its handling can be realised manually or by a standard robot equipped with a pick-and-place module. In this case, both solutions were realised, since the weight of the cavity does not exceed 12 kg for a 270 x 170 mm laminate.

To ensure that the thermal flux from or into the cavity begins simultaneously at the upper and the lower surface of the cavity, additional spring elements lift the cavity and create a narrow slot between compression toolings when the cavity is deposited. Contact between the tooling and the cavity is created not before the upper tooling applies pressure and the whole system is compressed (Figure 43).

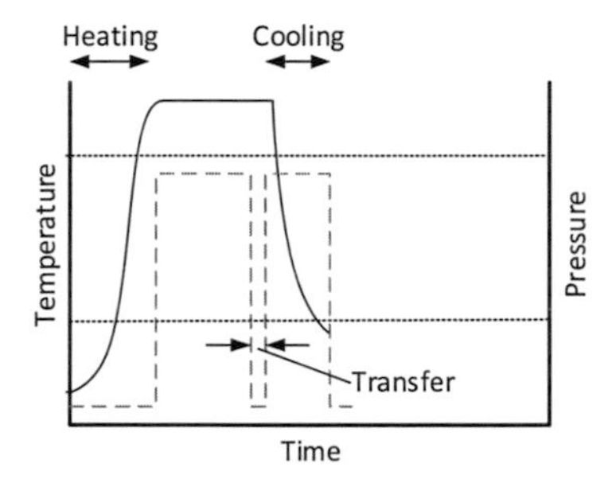

Figure 43 – Transfer stage in process

5.3 Energetic efficiency of Isoforming and variothermal thermoforming

Beside short cycle times for heating and cooling, the Isoforming process distinguishes in terms of its energetic efficiency. To prove this, a comparison of the required energy between Isoforming and classic variothermal compression moulding is drawn in Table 20. For the variothermal process, a cavity being equipped with electric cartridges and channels for cooling oil is considered. So, heat is provided electrically, and the cooling fluid is provided by a cooling system based on thermal oil. In this context the provided data for the energy consumption are based on data-sheet values.

Table 20 – Energy demand of variothermal – and Isoforming process

			Variothermal	Isoforming
Tooling parameters	tooling mass	[kg]	155	12
	thermal capacity - c_p	[J \kg K]	460	
	$T_{processing}$ - $T_{demoulding}$	[K]	250	
Heating of tooling	required energy	[kWh]	4.95	0.38
	energy costs	[EUR/kWh]	0.09	0.09
	costs	[EUR]	0.45	0.03
	fraction of overall process costs	[EUR]	67%	16%
Cooling of tooling	power of hydraulic pump (cooling unit)	[kW]	4	
	duration of cooling	[sec]	600	180
	energy cooling	[kWh]	0.67	-
	energy costs	[EUR]	0.06	-
	fraction of overall process costs		9%	0%
Energetic losses by convection	duration per cycle	[min]	45	25
	energy loss / area	[kW/m²]	12	12
	heated surface	[m²]	0.16	0.32
	convective energy loss	[kWh]	1.42	1.58
	cost of convection	[EUR]	0.13	0.14
	fraction of overall process costs		19%	64%
Energy for press	power press	[kW]	18	18
	duration / pressure generation	[sec]	10	10
	amount pressure generation / cycle	[1]	2	4
	duration / pressure rebuild	[sec]	2	2
	amount pressure rebuild / cycle	[1]	30	30
	duration operation of hydraulic motor	[sec]	80	100
	required energy	[kWh]	0.40	0.50
	costs for press operations	[EUR]	0.04	0.05
	fraction of overall process costs		5%	20%
Summary	costs	[EUR]	0.67 100%	0.22 33%

This estimation of the energetic demand shows a remarkable drop of required resources and a parallel drop of costs. However, with costs in dimensions below 1 EUR, the energetic cost will be far below the costs for the preforms or equipment as release agents respective films, for instance.

5.4 Conclusion – Isoforming process

This chapter introduced the Isoforming process and its characteristics. Contrary to classic variothermal processes, the variothermal fraction of the tooling is downsized to a minimum, so that for a simple laminate the tooling weight can be reduced from 155 to 12 kg (Figure 44). This simplifies handling, increases the heating/cooling rates into dimensions above 60 K/min and still enables temperatures up to 400°C or above. The developed shear-edge design furthermore reduces tolerance problems caused by the thermal expansion significantly.

In contrast to thermoforming of fully pre-impregnated fabric-based laminates, hybrid textiles allow complex fibre orientations. Joggles and local reinforcements by overlapping plies do not pose problems for the tooling, since they can be considered in mould and preform. Generally, special focus needs to be on the closing stroke if local thickness variations occur, due to locally different initial preform thicknesses and consequently locally longer strokes.

Costs for tooling material and machining are also reduced, since channels for a thermal fluid or special guiding elements can be saved. Extra costs for invest arise once by the stacked heating/cooling module and its thermal management.

Concerning process time, classic thermoforming with cycle times in the range of a minute is out of reach for the Isoforming process. Its cycle time focusses on high-performance components with specially designed fibre orientation and complex 2,5D-shape. The process is in competition with RTM or autoclave processes, where it is absolutely competitive with cycle times between 10 and 30 min for processing temperatures of 400°C. For those components, a previously preforming step is more laborious than for classic thermoforming, which must be considered in the overall process review. Beside hybrid textiles, preforms for Isoforming can be made of pre-impregnated tapes as well, which broadens the spectrum of application.

Instead of using hazardous fluids for thermal management of the tooling, electric cartridges provide heat for both heating and cooling. Hence, costs for investments like fluid-based temperature control systems are saved on the one hand. On the other hand, pressurised thermal oil with temperatures up to 400°C is prevented.

To conclude, the Isoforming process provides a suitable solution for processing hybrid textiles containing high-performance thermoplastics.

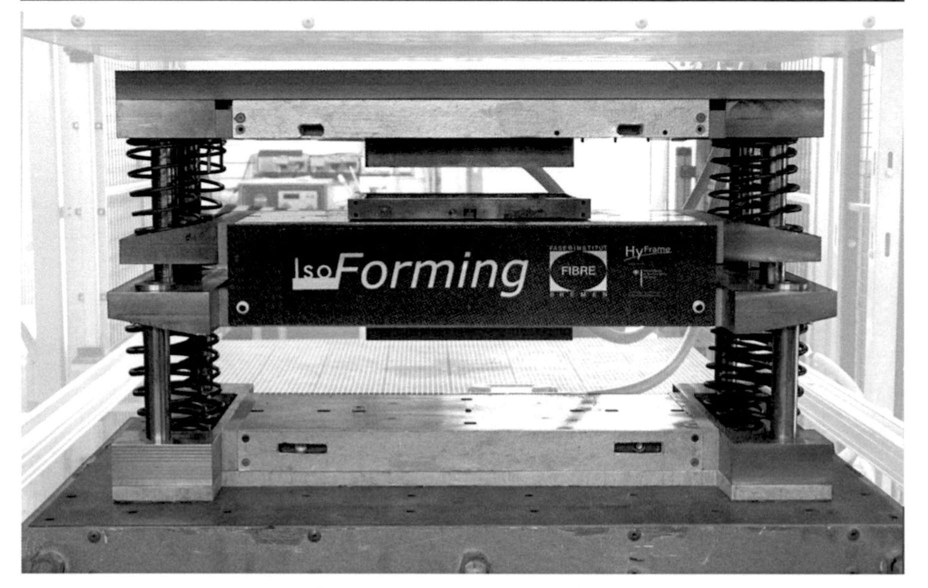

Figure 44 – Isoforming tooling

6 Model development for processing hybrid textiles

The following chapter discusses the development of a model for predicting the relevant processes during thermoforming of hybrid textiles. The model shall recommend parameters for thermoforming and is generated parallel to the manufacturing process. Consequently, both the model and the process depend on each other.

According to Advani [ADV10], a model contains six elements (listed below), which will be outlined in this chapter and in the following chapters:

1. Model or system boundary
2. Physical laws
3. Constitutive laws
4. Boundary conditions
5. Assumptions
6. Experimental validation

Different stages of thermoforming are illustrated and the significant effects are transferred into a numerical model. In general, the three different stages of heating, impregnation and solidification are modelled independently from each other, giving suitable recommendations for process parameters and sensitivities.

Table 21 gives an overview about the occurring phenomena during thermoforming. It comprises phenomena on three different scales: on macro-level, containing the whole textile; on meso-level, containing the roving level; and on micro-level, containing the interactions between single filaments and the matrix-system. Going further into detail, also the molecular level could be considered, which is neglected here. The further developed models will cover the phenomena marked with an "*".

Table 21 – Phenomena during Isoforming process with hybrid textiles (macro-, meso- and micro-levels)

	Macro-level	Meso-level	Micro-level
Heating	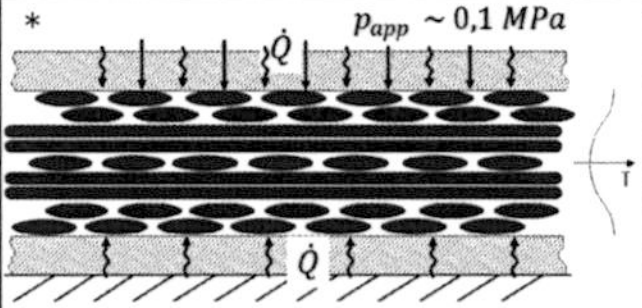 * $p_{app} \sim 0{,}1\ MPa$ $\dot{Q}$ T $\dot{Q}$	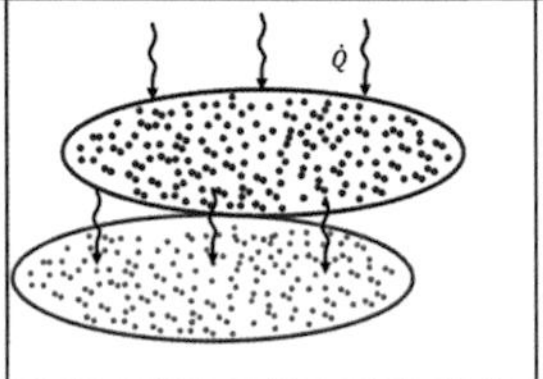 $\dot{Q}$	
Macro-Impreg. incl.	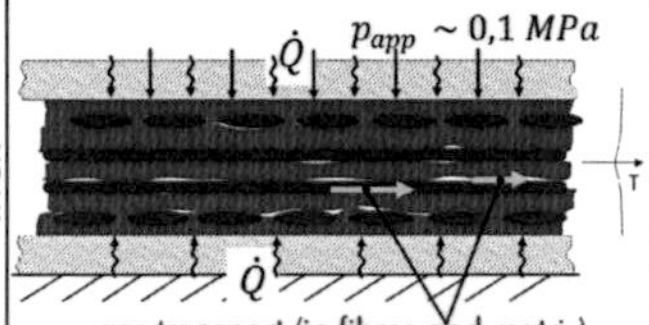 $\dot{Q}$ $p_{app} \sim 0{,}1\ MPa$ T $\dot{Q}$ gas transport (in fibres and matrix)	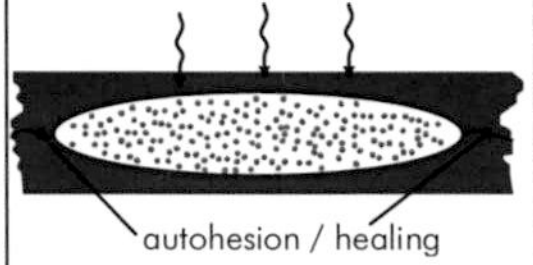 autohesion / healing	
Start meso-impregnation	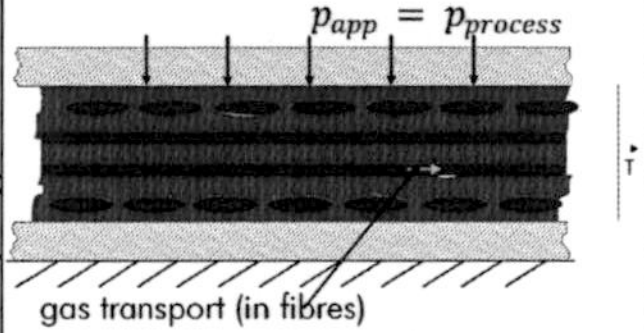 $p_{app} = p_{process}$ T gas transport (in fibres)	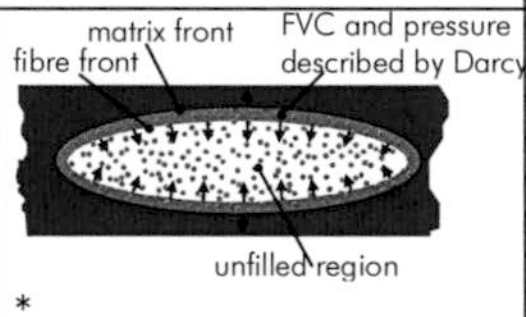 matrix front fibre front FVC and pressure described by Darcy unfilled region *	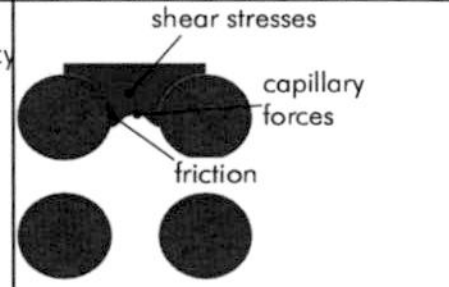 shear stresses capillary forces friction
Meos-impregnation	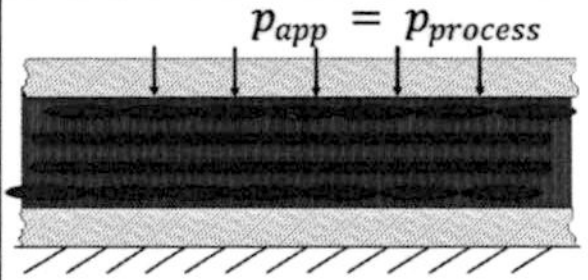 $p_{app} = p_{process}$	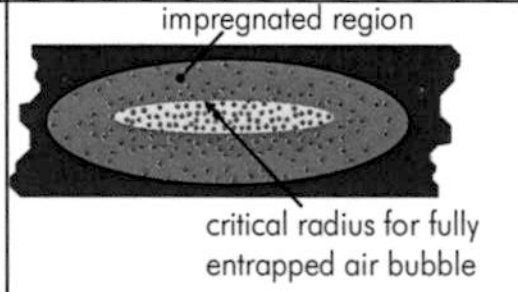 impregnated region critical radius for fully entrapped air bubble *	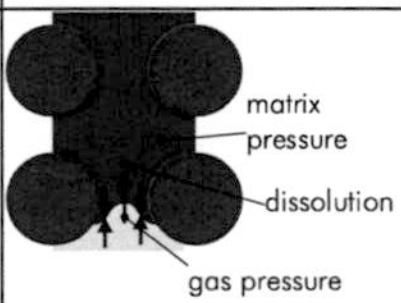 matrix pressure dissolution gas pressure
Transfer	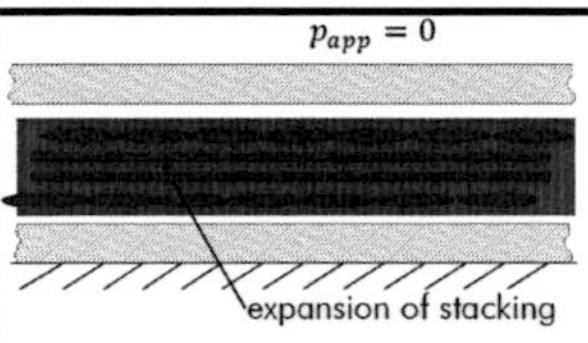 $p_{app} = 0$ expansion of stacking	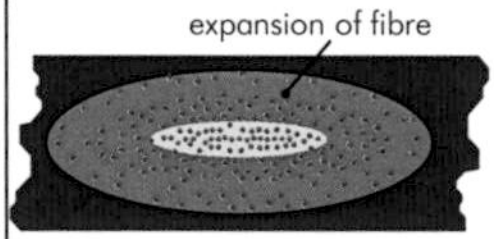 expansion of fibre	
Cooling / Solidification	$\dot{Q}$ $p_{app} = p_{process}$ T $\dot{Q}$ crystallisation $= f(\dot{T})$ therm. and cryst. induced shrinking *(without crystallisation)	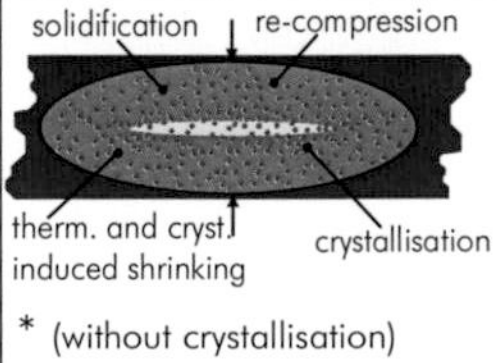 solidification re-compression therm. and cryst. induced shrinking crystallisation * (without crystallisation)	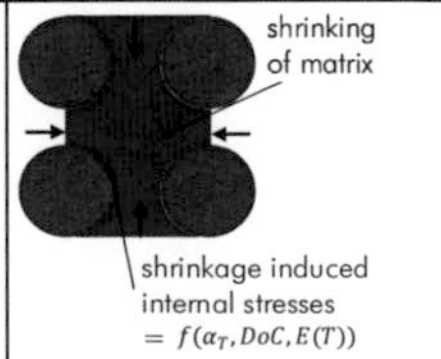 shrinking of matrix shrinkage induced internal stresses $= f(\alpha_T, DoC, E(T))$

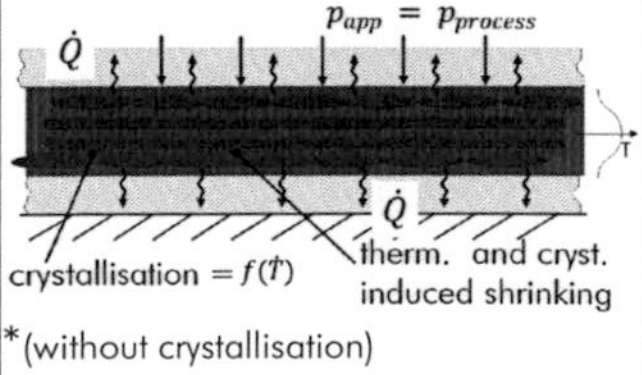

6.1 Heating

This chapter covers the modelling of the heating process of the tooling system and the non-impregnated hybrid preforms inside the hot cavity (Figure 45). On the one hand, being aware of the heating behaviour of the preforms allows to minimise durations of the heating phase. On the other hand, it gives information about the time when pressure can be applied for initiating a simultaneous start of impregnation in the whole preform. Consequently, the motivation of modelling is to know at which time the temperature at the centre of the preform is at the same level as at the preform surfaces. A prerequisite for modelling is the knowledge of the thermal conductivity and the single ply thickness of a preform in solid state. These data are acquired in section 4.2.

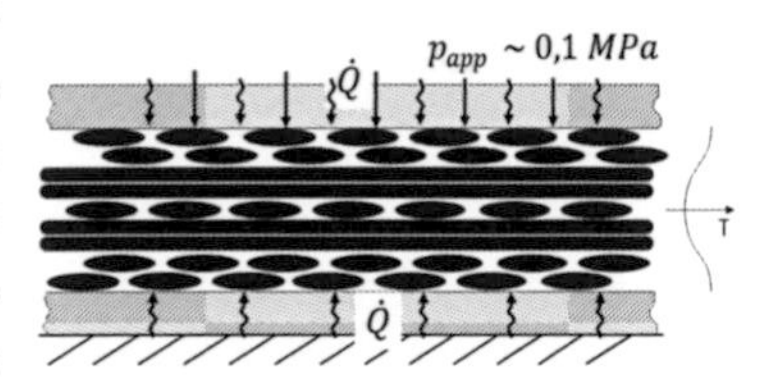

Figure 45 – Isoforming - heating

Literature and models

Other authors have already investigated the topic of heating thermoplastic prepregs or reinforcement textiles. Schaefer et al., for instance, analysed the heating characteristic of PA6-based prepreg material, especially focussing the thermal contact resistance [SCH16]. Fourier´s law describes the temperature distribution in the stacking:

$$\rho\, c_p \frac{\partial T}{\partial t} = \lambda \frac{\partial}{\partial x}\left(\frac{\partial T}{\partial x}\right) + \dot{q}. \tag{6.1}$$

However, for a one-dimensional heat transfer problem, the equation can be simplified. The last term $\dot{q}$ describes the internal or external heat sinks or sources, which is neglected for heating of the preform. The required enthalpy during melting is considered by the temperature-dependent thermal heat capacity.

Modelling of temperature distribution of tooling and preform

Modelling of the transient temperature distribution of the assembly shall deliver process constraints regarding feasible heating rates and an estimate for the homogeneity of the temperature distribution. In order to investigate if heating of the tooling or heating of the textile is faster, two independent heating models are developed. The first model considers only heating of the hybrid textiles, neglecting the time required for heating of the cavity.

In the second model, the overall setup including tooling and preform is considered concerning its heating characteristic.

A conservative approach is followed, if only thermal conduction into the preform is considered without any mass transport, which could be caused by the molten polymer for temperatures above T_m. An appropriate simplification is to model the preform stacking as a continuum. Taking an instantaneous temperature jump at the tooling to processing temperature into account, the corresponding temperature inside the preform is illustrated in Figure 46. The results are determined using a one-dimensional MATLAB code considering equation (6.1).

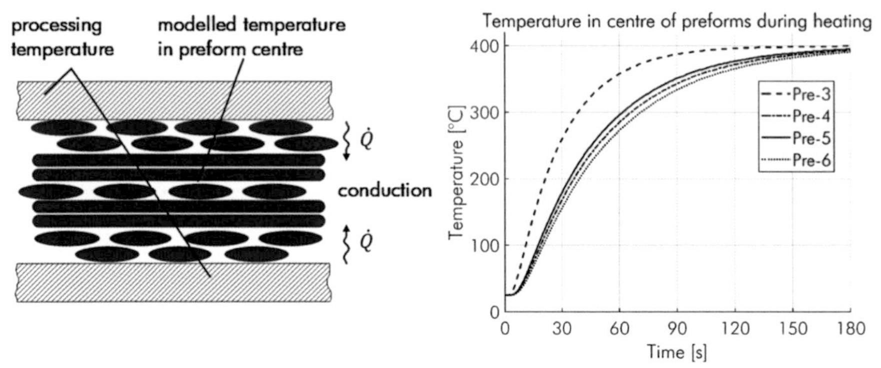

Figure 46 – Simplified model for heating a hybrid textile for the centre temperature during heating at 0.1 MPa for instantiation with a surface temperature of 400°C

The results clearly indicate the influence of the thermal conductivity on the temperature gradient. While the centre temperature in Pre-3 crosses 380°C already after 80 s, it takes about 150 s in Pre-4, Pre-5 and Pre-6. This leads to the recommendation to wait 1–3 min from the time reaching a tooling temperature of 400°C until the consolidation process and the corresponding application of pressure shall begin. It must be addressed that the temperature increase to 400°C of the cavity is not instantaneous in reality.

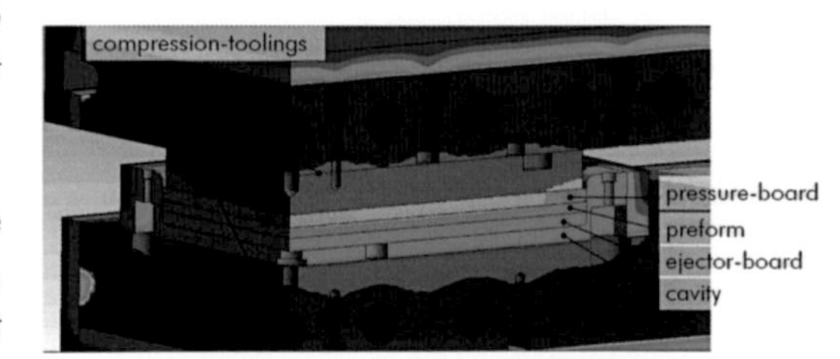

Figure 47 - FE-model for heating

In the next step consequently, the whole system including cavity, compression toolings and heat sinks is considered for modelling the heating characteristic (Figure 47). The analysis is performed by a finite element analysis using the software ANSYS. All relevant parameters are listed in Table 22. Here, the parameters for the thermal contact resistance and the heat transfer coefficient for natural convections are based on assumptions. Regarding the thermal contact resistance in the interfaces of the mould components (compression tooling – cavity – pressure board), its value is primarily influenced by the acting contact pressure, the material of the counterparts and the surface roughness of the contact areas. Literature provides values between 0,6 and 2,5 m^2K/W (at 0 and 1 MPa) and 0,07 and 0,4 m^2K/W (at 10 MPa) for the resistance between solids of steel [INC96]. Since the surfaces of the toolings are polished, a value closer to the smaller resistance value is chosen and extrapolated to contact pressures of 2 MPa. For cooling, caused by the ambient air, only natural convection is considered. Here, an estimation of convection at a vertical plate delivers a heat transfer coefficient of 10,7 W/m^2K between tooling and environment, which is also considered by Brauner [BRA13]. Material transport after melting is neglected as well as thermal radiation, since its value is less than 5% of the convective heat flux, leading to a conservative approach.

Table 22 – Parameters for transient thermal simulation

			Source
Thermal contact resistance steel-steel	[m^2K/W]	8,4 e-04	[INC96]
Tooling components (steel 1.2312)			
Initial temperature	[°C]	400 / 360	
Thermal conductivity	[W/mK]	34,5	[SAU18]
Thermal capacity	[J/kgK]	460	[SAU18]
Density	[g/cm^3]	7,85	[SAU18]
Cavity (steel 1.2343)			
Initial temperature	[°C]	30	
Thermal conductivity	[W/mK]	25	[SAU18]
Thermal capacity	[J/kgK]	460	[SAU18]
Density	[g/cm^3]	7,80	[SAU18]
Composite (unconsolidated state)			
Thermal diffusivity	[mm^2/s]	scalar	Table 16
Insulation (AGK K-Therm AS 550)			
Thermal conductivity	[W/mK]	0,37	[AGK18]
Thermal capacity	[J/kgK]	1000	assumption
Density	[g/cm^3]	1,8	[AGK18]
Ambient conditions			
Ambient temperature	[°C]	40	
Heat transfer coeff.– natural convection	[W/m^2K]	10,7	

Analysing the heating trajectory of the different preform configurations points up, that all specifications show a similar heating characteristic despite their different thermal diffusivity. An exemplary heating curve for the upper-, centre- and bottom-positions of a Pre-3 preform and for the centre-position of Pre-4 is shown in Figure 48. The left diagram illustrates a slow heating of the cavity surfaces, which is the limiting factor for the heating rate. Consequently, to increase the heating rates inside the preform, the thermal conductivity of the tooling system needs to be improved as a first step. The conductivity of the preform itself only has secondary influence. A further result is a homogeneous temperature distribution along the preform thickness. After approximately 7 min a processing temperature of ~380°C is reached, so that pressure can be applied. For Pre-3 with a high fibre volume content (~60%) and Pre-4 with a smaller fraction (~50%), no significant distinction can be determined. Hence, the thermal conductivity and the corresponding fibre volume content only have secondary influence on the heating time.

On the right-hand side of Figure 48, the heating characteristic regarding different preform thicknesses is shown. Here, the model illustrates a longer response time until temperature increases in the preform centre. After 7 min though, temperatures inside 4- and 10-mm-thick preforms converge.

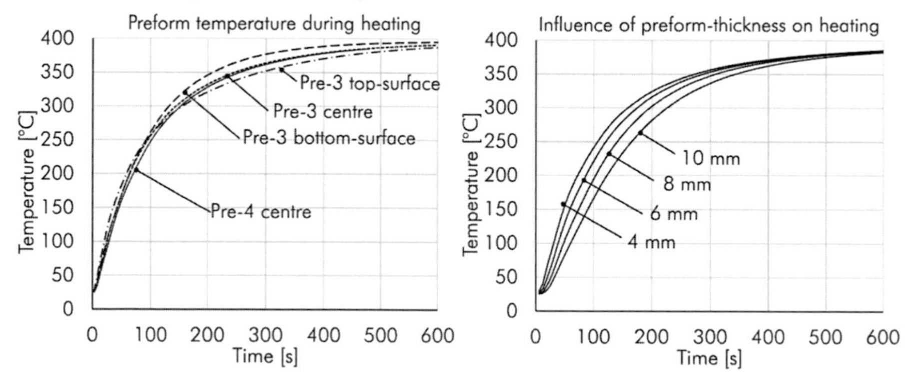

Figure 48 – Modelled temperature distribution in preform during heating (left); temperature in Pre-3 centre for different preform thicknesses (right)

To recapitulate the results of this chapter, the duration of the heating stage of the Isoforming process is predominantly governed by the heating of the steel components (cavity, pressure and ejector board). Therefore, for increasing the process speed, a reduction of the thermal contact resistance between those elements is the most promising approach. A suitable method is the reduction of the surface roughness by polishing. Neither the thermal conductivity nor the thickness of the preform shows a significant influence on the heating rate. For process temperatures of 400°C, the time required for heating is in the range of 6-8 min, which will be validated in the experiments.

6.2 Impregnation

The objective of impregnation is to create a void-free structure only containing homogeneously distributed fibres inside a polymer matrix, which consolidates the composite by means of adhesion. Prerequisites for an appropriate impregnation are among others:

- an externally applied pressure
- sufficient low melting viscosity of the polymer due to a respective temperature
- sufficient out-of-plane permeability of textile (on meso- and micro-scales)
- free channels inside the preform to squeeze out residual air
- homogeneous in-plane pressure distribution in stacking to prevent lateral squeeze flow of fibre–matrix structure.

In the following, the different phenomena of impregnation are described more closely. A focus is on into the process of compaction, macro- and meso-impregnation. Afterwards, the development of the impregnation model and a parameter analysis are performed.

6.2.1 Macro-impregnation including autohesion / healing

Compaction

Prior impregnation, an intimate contact between the single plies needs to be assured by compaction to reduce free space to a minimum and to squeeze out residual air. Spatial gaps are usually a feature of unevenly distributed fibres and shall of course be reduced.

Macro-impregnation

After the phase of first compaction, macro-impregnation begins after melting of the polymer at applied pressure (Figure 49). In this phase, the temperature distribution still can be inhomogeneous across the thickness and the corresponding start of melting might be locally varying. However, if squeeze flow along a pressure gradient shall be prevented, the temperature distribution across the preform in in-plane and out-of-plane directions should be as homogeneous as possible.

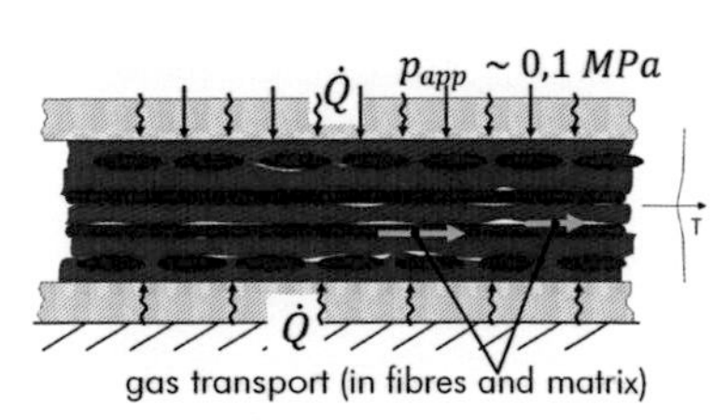

Figure 49 – Isoforming – macro impregnation

Permeability is closely linked to the characteristic diameter of the porous medium. Comparing the radius of a fibre bundle to that of a filament and respecting the corresponding equations in section 4.5 delivers a difference of permeability of five or more decades. Therefore, the first process of impregnation takes place at macro-level around fibre bundles or rovings. This procedure was also detected in several experimental works, for example, Wakeman, Long or Ehleben [WAK98, LON01, EHL01]. The fluid flows around the bundle and coalesces with the opponent flow.

Autohesion / Healing

The next stage, when adjacent matrix interfaces penetrate each other by intermolecular diffusion, forms the matrix network [VOY63]. Interfaces between the polymer flows disappear in favour of a continuous matrix system, which is called autohesion or healing. Approaches for modelling the degree of healing D_h are demonstrated exemplarily by Gutowski or Yang [GUT91, YAN02]:

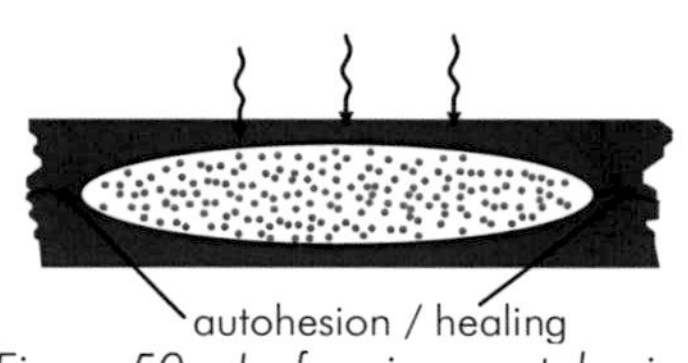

Figure 50 – Isoforming - autohesion

$$D_h = \left(\frac{t}{t_r}\right)^{1/4}.$$

Stokes-Griffin and Bouwman report reptation times t_r for PEEK in dimensions below 1 s for temperatures in dimensions of 400°C, which is why autohesion is a negligible phenomenon here and will not be considered in further modelling [STO16, BOW16].

6.2.2 Meso-impregnation

Start of meso-impregnation

During meso-impregnation, the matrix front starts penetrating the roving. In the beginning of meso-impregnation, driving forces are again the externally applied pressure and an additionally capillary pressure which supports the flow.

Progress of meso-impregnation

The segment of the partially impregnated roving can be modelled according to Darcy´s law. Following Terzaghi's principle, the externally applied pressure load is transferred from the liquid polymer into the textile. At the flow front, the textile absorbs the whole load. Again, remaining air is squeezed out, primarily through free channels inside the roving. However, from a certain degree of impregnation, air can be fully entrapped, which leads to a formation of voids (Figure 51).

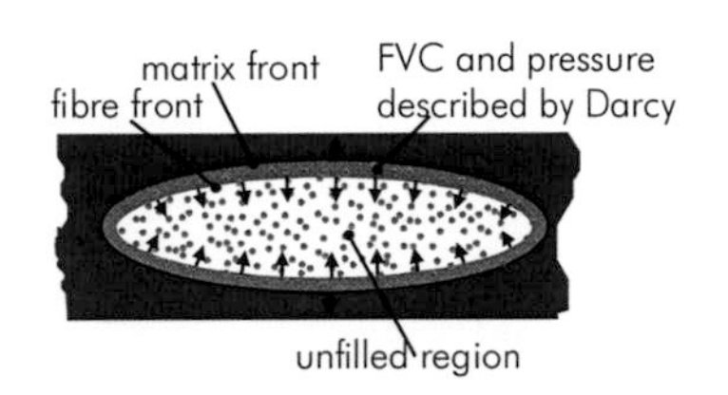

Figure 51 – Isoforming - meso-impregnation

Formation of voids

Up from a certain critical radius inside the partially impregnated rovings, gas bubbles are entirely entrapped. From this point on, the pressure inside the air bubble counteracts the fluid pressure according to the ideal gas law. During compression, the pressure inside the entrapped air increases until it equals the external isostatic pressure inside the molten polymer. Beside pressure, the surface tension of the fluid and its viscosity influence the void size. If not fully compressed or diffused into the polymer, it will form a void after solidification [Li15].

Several authors describe the process of migration of voids during re-consolidation after a pressure drop at processing temperatures. This topic is especially of interest for welding and thermoforming of pre-consolidated laminates, since a de-consolidation supports void growth [Shi16, Lu04]. These voids can be compressed in a subsequent compression stage, so as here. The time required for compressing the voids until total migration was found to be in fractions of a second. Consequently, it can be assumed that the transfer process between the hot and the cold toolings and the accompanied pressure drop can provoke void growth. This however is reversible in a subsequent compression stage during the phase of solidification.

For high-performance thermoplastic composites from hybrid textiles, entrapment of air is the most likely process for the formation of voids. Other reasons for entrapped gas, as condensation for instance, are less significant due to the low absorption of humidity of PEEK, which is in the range of 0,5%. Anyway, to minimise this effect, the preform shall be dried prior consolidation.

Solubility of gases in polymers

Generally, diffusion of gases into polymers can be described by Fick´s laws, which express the diffusion as a result of different gas concentrations and the diffusion rate in terms of a diffusion coefficient [FIC55]. Furthermore, Henry´s law describes the correlation between the applied pressure and the solubility of gases [HEN03]. An implemented description of the solubility of gases in polymers was perfomed by Nilsson et al., who improved the non-equilibrium lattice fluid theory (NELF). They validated the model with Victrex PEEK 450G at room temperature. An almost linear dependency between the saturation concentration of nitrogen and pressure was examined. For pressures at 1 MPa, 1 cm^3 of PEEK could store approximately 1 cm^3 of nitrogen, respectively, at room temperature and ambient pressure [NIL13].

Note that only the amorphous fraction of the polymer can solute gas. Consequently, the saturation concentration decreases with increasing degree of crystallinity. For solid state, the saturation concentration was shown to be independent of temperature. For temperatures above the melting region, no data are available; however, for other polymers (PP, PS), the respective saturation concentration increases at increasing temperature of the polymer fluid [ARE04]. Thus, considering a saturation concentration of 1 cm^3 of nitrogen per 1 MPa pressure is a conservative approach.

End of meso-impregnation

Two phenomena can finish meso-impregnation. The desirable condition is, if the component is fully impregnated and the flow fronts coalesce inside the rovings. In case of entrapped air on the other hand, the flow front penetrates until an equilibrium between the driving pressure and the gas pressure is achieved. This results in voids remaining inside the laminate.

6.2.3 Modelling

For modelling the phenomena mentioned above, physical and constitutive laws are considered. This chapter describes those laws and aligns them in a superior model to give a comprehensive overview about impregnation.

Assumptions

Below the model for impregnation is described, acting on several simplifications and homogenisations of material and process parameters, as well as occurring phenomena. Concerning temperature, a homogeneous distribution is assumed, where no further heat is generated by shear flow or submitted from external heat sources. Furthermore, a constant pressure in z-direction is adopted. To simplify the model, only flow in z-direction is proposed. At last, impregnation stops whether when flow the front has crossed the entire flow path or when pressure equilibrium arises between fluid pressure and entrapped gas bubbles.

Physical laws

The description of flow is generally described by the equations of conservation of mass and conservation of momentum, which are described below.

Conservation of mass (solid and liquid phase)

The following formulation of the equation for the conservation of mass presupposes an incompressibility of fibres and fluid. Furthermore, changes of velocity or stresses inside one control volume can be assumed to be negligible. The general form of the continuity equation is:

$$\frac{\partial \rho}{\partial t} + \frac{\partial \rho u_i}{\partial x_i} = 0\,.$$

Considering the motions of fluid and textile yields to the adapted descriptions:

$$\frac{\partial \varphi_F}{\partial t} + \frac{\partial}{\partial z}(\varphi_F u_F) = 0 \qquad (6.2)$$

$$-\frac{\partial \varphi_F}{\partial t} + \frac{\partial\big((1-\varphi_F)u_M\big)}{\partial z} = 0. \qquad (6.3)$$

Conservation of momentum

The conservation of momentum of one-dimensional flow in its general form, neglecting gravitational effects, yields:

$$\frac{\partial u_z}{\partial t} + u_z \frac{\partial u_z}{\partial z} = -\frac{1}{\rho}\frac{\partial p}{\partial z} + \frac{\mu}{\rho}\frac{\partial^2 u_z}{\partial z^2}\,.$$

For modelling, a simplified form of the Navier–Stokes equations is necessary, which will be described subsequently by Darcy´s law.

Constitutive laws

Force equilibrium between fibre network and matrix phase

An equilibrium of forces is assumed for the preform–matrix network. This means, that the sum of pressure acting on the textile and onto the fluid remains equal, so that a rising pressure taken by the textile leads to a reduced fluid pressure. This context is also known as Terzaghi´s principle:

$$\frac{dp_M}{dz} + \frac{d\sigma}{dz} = 0 . \tag{6.4}$$

Here p_M defines the pressure inside the matrix system, while σ represents the stress absorbed by the textile.

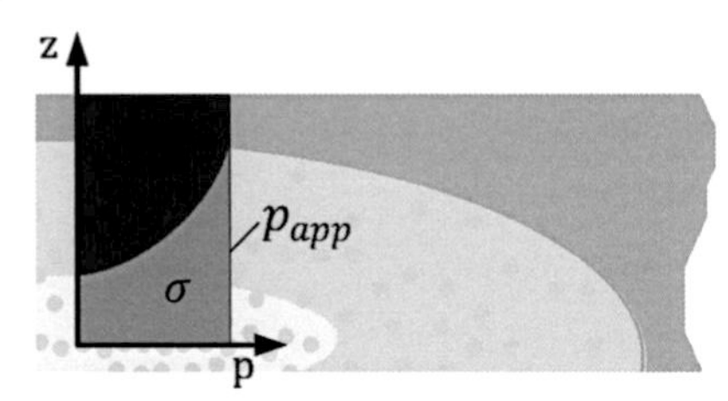

Figure 52 – Terzaghi´s principle for impregnated regions on meso-level

Darcy´s law

Darcy´s law can describe the progress of penetration through porous media. It is a simplification of the Navier–Stokes equations and thus representing the conservation of momentum. It allows a simplified calculation of the flow instead of using the entire momentum equations, which would require a description of the structure of the preform on micro-scale. Despite exponentially growing computation capabilities, this detailed description of the current geometric state is not feasible and would not lead to further essential information since normally process parameters on meso-scale describe the problem sufficiently. The model describes the flow rate being influenced by a pressure gradient, the fluid viscosity, a material-dependent parameter representing the permeability and time:

$$\frac{dz}{dt} = \frac{K}{\eta} \frac{dp_M}{dz} . \tag{6.5}$$

This form of Darcy´s law is applicable if viscous forces dominate over shear forces inside the flow, which is the case for a Reynolds number smaller than Re < 1. For current setups in macro-, meso- and micro-scales, the corresponding Reynolds numbers are far below unity for all scales, which allows the application of Darcy´s law:

$$Re = \frac{\rho u L}{\eta} = \frac{shear\ forces}{viscous\ forces} \ll 10^{-6}\,.$$

Without Consideration of fibre-bed elasticity

If fibre relaxation for impregnated regions is neglected, so that only the distance covered by the polymer flow is calculated, Darcy´s law simplifies to:

$$\Delta t = \frac{\eta \cdot z^2}{2 \cdot K \cdot \Delta p_M}\,. \tag{6.6}$$

Here especially the permeability causes uncertainties, since it is assumed to be independent from the applied pressure. Further improvement can be achieved by modelling the approaches described in section 4.5. Taking the Gutowski model into account leads to the following expression describing the impregnation:

$$\Delta t = \frac{2\,\eta\,k_z\left(\frac{\varphi_{max}}{\varphi_F}+1\right)z^2}{r_f^2\left(\sqrt{\frac{\varphi_{max}}{\varphi_F}}-1\right)^3 (p_{app}-p_{amb})}\,.$$

Note that φ_F is linked with the applied pressure. This analytical equation already gives a convenient overview about the governing parameters of impregnation. Especially, the required flow length and the pressure dependent fibre volume content are of great importance.

With consideration of fibre bed elasticity

If fibre relaxation is considered as illustrated in Figure 53, Darcy´s law as written in equation (6.5) needs to be supplemented by motion of fibres. Furthermore, considering the fibre volume content yields:

$$u_M - u_F = \frac{K}{\eta(1-\varphi_F)}\frac{dp_M}{dz}\,. \tag{6.7}$$

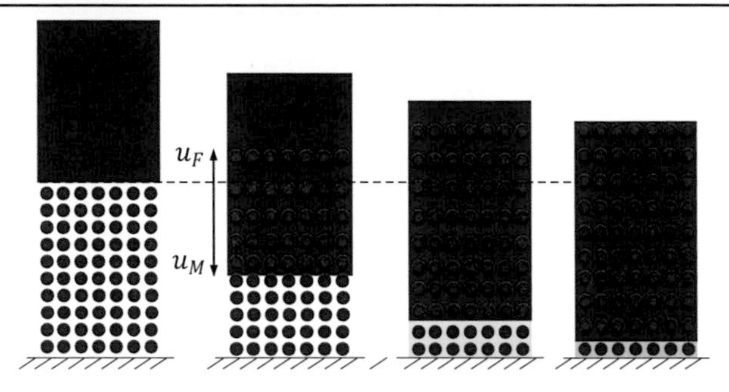

Figure 53 – Modelled stages of impregnation

Here again only flow in the out-of plane direction is considered. It is assumed that textile relaxation immediately begins with passing flow front. The compressive strain acting on the textile decreases towards the textile front, following the stress–strain curve for unloading (see chapter 4.4). This can be exemplified by illustrating the textile as a series of infinitely thin springs. As the liquid flows past these, it exerts a viscous drag force on the objects, which pushes these forward to the infiltration front.

The model and the further numerical approach for solution are mainly based on works of Sommer, Michaud, Jespersen and Wysocki [SOM92, MIC97, MIC01, WYS05, JES08].

Permeability

Concerning the textile, a homogeneous distribution of filaments and fibres in in-plane and out-of-plane directions is presupposed. A one-dimensional Gebart-model with quadratic filament arrangement is considered as described in section 4.5 and especially in equation (4.8):

$$K_{Geb,\perp} = c_1 \left(\sqrt{\frac{\varphi_{F,max}}{\varphi_F}} - 1 \right)^{\frac{5}{2}} r_f^2 \, .$$

Since the most relevant region for impregnation of hybrid textiles is on roving level, the model only considers the meso-scale. This takes the filament radius into account and the fibre volume content inside a roving.

Viscosity

Incompressibility of filaments and fluid is a further assumption, which was shown to be valid before. Moreover, shear thinning of the polymer fluid is neglected. Various authors modelled shear rates at macro-, meso- and micro-levels during impregnation. Haffner et al. used finite element analysis to estimate resulting shear rates on microscopic scale for pultrusion of PP and glass fibres resulting in shear rates below 0,3 s^{-1} [HAF98]. Ehleben showed that for impregnation of glass fibres with PA6, shear thinning becomes relevant above 1 mm/min, which is far below the expected velocities here [EHL01].

Since a Newtonian fluid is assumed, the only dependency concerning viscosity is temperature, which remains constant throughout impregnation. Consequently, the simplified Arrhenius relationship with parameters for low shear rates is considered as described in section 3.1:

$$\eta(\vartheta) = \eta_0(T_0) \cdot e^{\frac{E_0}{R}\left(\frac{1}{T}-\frac{1}{T_0}\right)} .$$

Formation of voids

The formation of voids is essential for further laminate properties. Here it is modelled on meso-level. Compared to meso-scale, permeability on macro-scale is significantly higher, which normally prevents the creation of voids on macro-scale.

Although diffusion of voids into the polymer is a probable effect, it is not considered here due to an uncertain knowledge about the dynamics of the process and its restraints. This is a conservative assumption, which will not lead to overestimated void contents.

Compression effects a pressure rise inside the void acting against the impregnation. The characteristics of the pressure inside the void are described by the gas law for ideal gases:

$$p_{Gas}\, V_{Gas} = n\, R\, T_{Gas} \,. \tag{6.8}$$

Here V_{Gas} and T_{Gas} represent the volume respective the temperature of the void, while n and R are the number of moles and the ideal gas constant. The model assumes that the voids are fully entrapped from a certain impregnation length z_c. This approach already has been verified on meso-level for entrapped air inside rovings and is further used here [BER99]. The determination of this specific length z_c is a challenge to choose. According to the gas law, the void pressure can be described as:

$$\begin{gathered} p_{Gas} = 0; \quad (z \leq z_c) \\ p_{Gas} = p_{amb} \cdot \left(\frac{z_c \cdot \left(1 - \varphi_F(z = z_c)\right)}{z \cdot (1 - \varphi_F)} \right); \quad (z > z_c) \,. \end{gathered} \tag{6.9}$$

By this, the calculated pressure will be considered in the model as counteracting impregnation. The critical impregnation length is defined as the fraction of the original thickness of the unimpregnated textile:

$$z_c = \kappa \cdot th_{CF}\,; \quad \text{with } 0 \leq \kappa \leq 1\,. \tag{6.10}$$

According to CT scans with corresponding measurements of voids, κ should be chosen close to one, since experiments demonstrated voids to be in magnitudes of 2 mm in lateral and maximum 0,5 mm in the out-of-plane direction [KOE18].

Capillary pressure

During impregnation, an instantaneous pressure drop of the preform after infiltration arises, due to the difference of solid/air and liquid/solid interfacial energies. This pressure drop contributes to impregnation and is called capillary pressure. For modelling impregnation, it is necessary to estimate the contribution of the capillary pressure to the total driving pressure to clarify if it is considerable.

Literature

A common way for modelling capillary forces is approached by Connor, who investigated macroscopic capillary pressure and microscopic interparticle forces and their effect on impregnation [CON95]. Since the equation is directly influenced by the reciprocal radius of the porous medium, only the micro-level is considered:

$$\Delta p_{cap} = \frac{2 \cdot \gamma_{se}}{r_f} \cdot \frac{\varphi_{micro}}{1 - \varphi_{micro}} \cdot \cos\theta\,. \tag{6.11}$$

Result

Beside the fibre radius, also the surface tension and the contact angle θ of a polymer droplet on carbon fibre take influence on capillary forces. For polymers, the range of surface energies γ_{se} is between $10 \cdot 10^{-3}$ and $50 \cdot 10^{-3}$ N/m [MIC01, HA97]. Referring to Ebnesajjad et al., values for PEEK are in dimensions of $42 \cdot 10^{-3}$ N/m [EBN13]. Since no reliable data for the contact angle can be found, a conservative value of 10° is considered. Considering the equation above with a maximum fibre volume content of 75% shows that even for conservative parameters the capillary pressure does not exceed 0,07 MPa. This value is far below the externally applied pressures and can thus be neglected.

6.2.4 Numerical simulation

In order to capture all constitutive equations and physical laws mentioned above, a numerical simulation is developed. In particular, it consists of a Boltzmann transformation to create two similarity variables, where one of them combines time and impregnation length, while the other represents a non-dimensional flow rate. This procedure was first applied by Sommer and further enhanced by Michaud and Jespersen [SOM92, MIC97, MIC01, JES08]. Their approaches are adapted in this work.

The rewriting of the equations for the conservation of mass and momentum is simplified by the similarity variables. The system of partial differential equations is transformed to an ordinary differential equation system. For solving, boundary conditions and initial conditions are arbitrarily chosen. An iterative solution delivers all state variables which are re-transformed into physical parameters.

The distinctive mathematical equations and their derivation are illustrated in the appendix. One ordinary differential equation describes Darcy´s equation for flow in porous media, which delivers information about the motion of fibres and fluid. Two ordinary differential equations enable the conservation of mass for fibres and fluid. The implementation is realised in a MATLAB code.

For iteration, a tailored scheme was developed for adjusting both values Ψ and s independently from each other. The sequence of the model is illustrated in Figure 54.

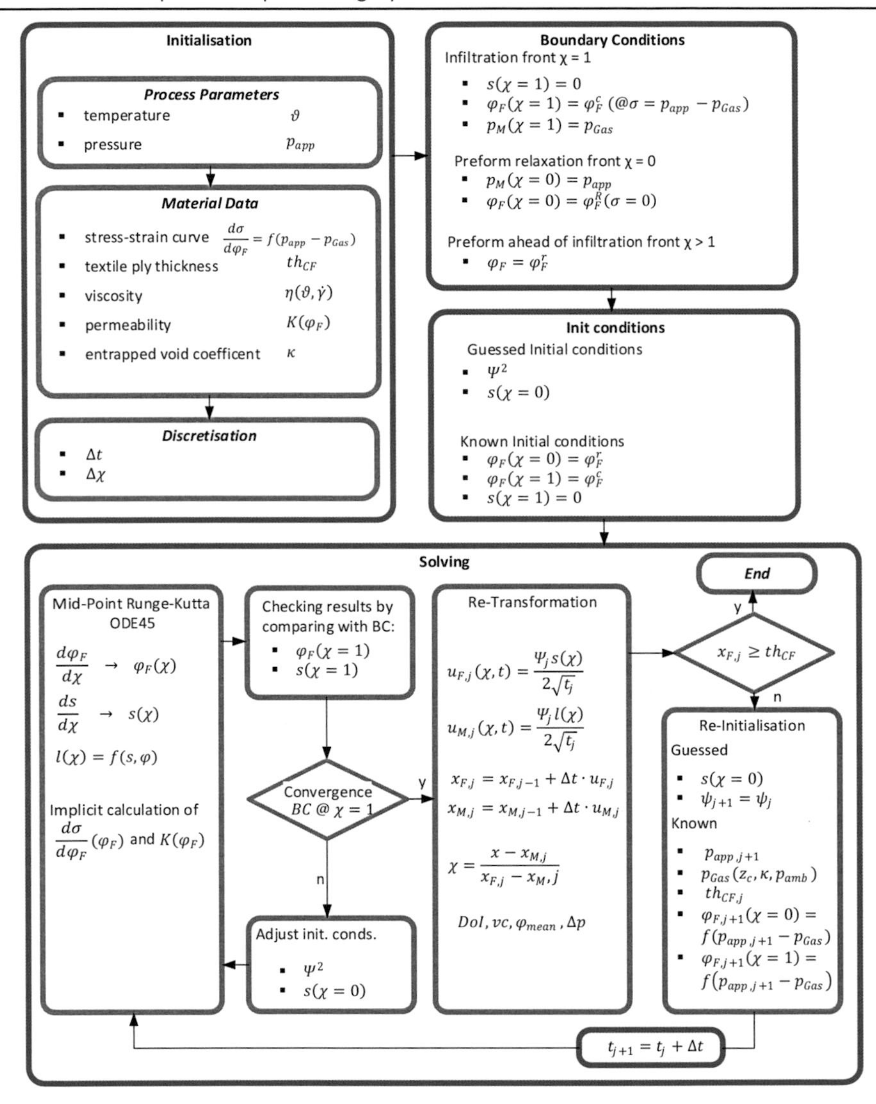

Figure 54 – Sequence of the impregnation model

As output, the code delivers the local fibre volume content and matrix pressure, depending on the permeability and compression characteristics. Furthermore, the velocities of the flow and the textile are expressed together with an overview chart about the current state of impregnation. An example of the temporal results is illustrated in Figure 55.

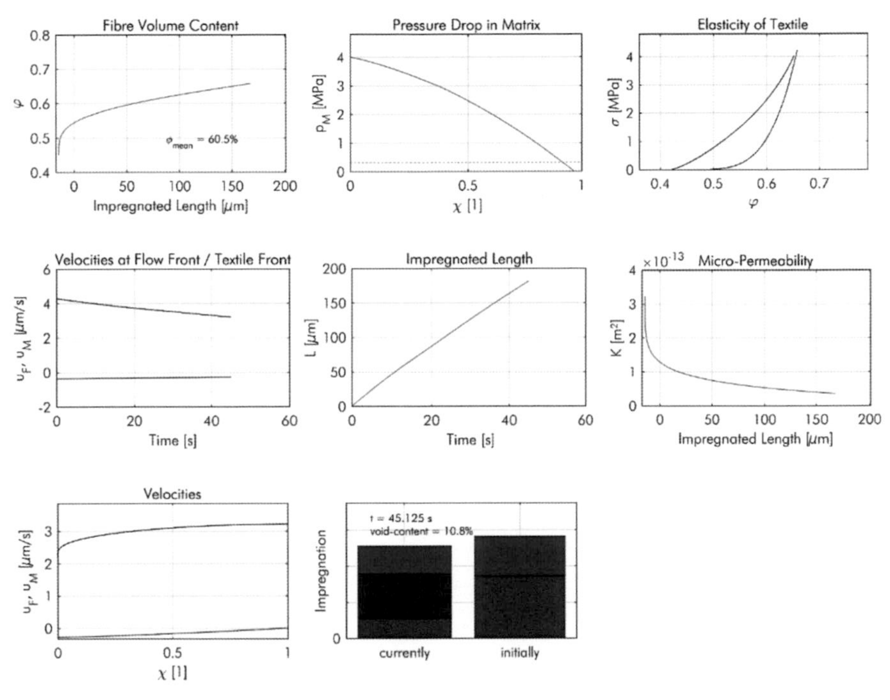

Figure 55 – Exemplary output data from impregnation model

6.2.5 Results and sensitivity analysis

The developed model shall help to identify the impact of the relevant process parameters regarding shortest impregnation times and void-free laminates. This objective is pursued in the following by adapting the applied pressure and the impregnation temperature and by considering the four different hybrid textiles, which have been introduced in chapter 4. Note that no substantial difference between the two investigated polymers PEEK 151G and the low-melt type AE250^{TM}-polymer is expected, since the only relevant polymer parameter here for impregnation is viscosity, which is comparable for both at their processing temperature.

In this context, the calculated absolute values must be taken with caution. As described above, several simplifications are made in the sub-models, which result in uncertainties of the absolute values. However, the integrated model considers the most considerable impregnation effects and can consequently help evaluate the sensitivity of impregnation with respect to process and material parameters.

The results presented below show a significant deviation between modelled and measured impregnation time. While experimentally identified required impregnation times are in dimensions of minutes, the modelled required durations are in dimensions of seconds. Therefore, the modelled time is shown as a relative value subsequently.

Textile parameters – Influence of impregnation length

Concerning the different hybrid textiles, they only distinguish in terms of the flow path length and their compression behaviour in the developed impregnation model. All further properties as local variations of the fibre distribution, scatter of the flow path length or the size of the thermoplastic fibres, for instance, are not considered in the model. Figure 56 shows the required time for impregnation with respect to the flow path length of the four different textiles. For each textile configuration, the required impregnation duration is illustrated at applied pressures of 4 MPa.

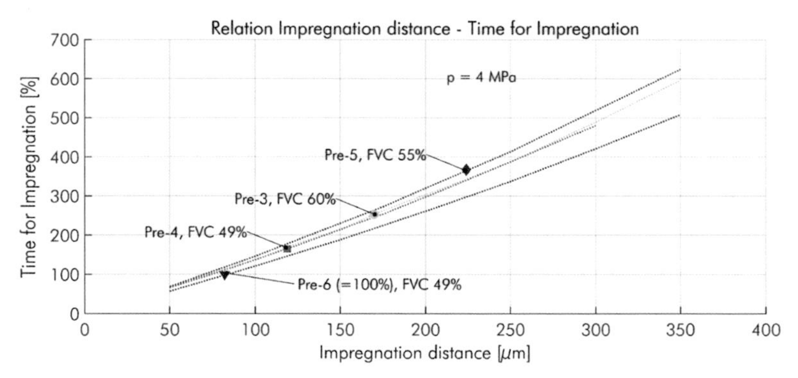

Figure 56 – Modelled resulting time for impregnation with respect to flow path length

Note that the modelled absolute values for the required impregnation time are too small compared to the experimentally investigated values, which will be presented in section 7.2.1. A reason for this is the inhomogeneity of the fibre–matrix distribution, which makes it more difficult to model the stress distribution between fibre and matrix. During impregnation a considerable amount of stress might be transferred by the thermoplastic fluid on macro-scale. This underlines the necessity to consider a multi-scale impregnation model which also takes macro-impregnation into account.

The results indicate an almost linear relation between impregnation time and impregnation distance. In contrast to the simplified interpretation of the impregnation time in equation (6.6), the model which takes fibre-bed elasticity into account indicates an almost linear (not quadratic) relation between impregnation time and flow path length, which is caused by the different stress-FVC relationships of the textiles.

Process parameters – Influence of pressure on time for impregnation

Regarding the relation between applied pressure and resulting required time for impregnation, no clear trend can be manifested, as shown in Figure 57. Three out of four textiles show a slightly faster impregnation with increasing pressure, while Pre-5 requires a minimum time at pressures in regions of 2 MPa. The other textiles require between 13% and 18% shorter impregnation times, if pressure is increased from 2 to 5 MPa. Thus, increasing the applied pressure only has a small effect on impregnation time.

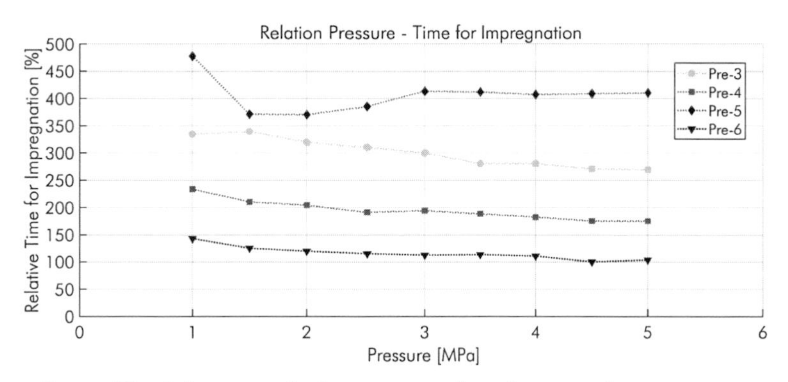

Figure 57 – Relation applied pressure and resulting time for impregnation

Process parameters – Influence of pressure on void content

In contrast to the impregnation time, the model indicates a significant influence of the applied pressure on the resulting porosity. Figure 58 demonstrates this relation. Generally, increasing the pressure effectuates a reduced void content. For a void content below 2%, the model proposes applied pressures above 2 MPa. Above 4 MPa, void contents below 1% are attainable according to the model. An interesting outcome are the comparable values of the void content for all different textiles despite their distinguishing flow path lengths. The model consequently proposes that the initial size of entrapped air has a negligible effect on the resulting void size. A presumption however is that an increasing flow path length on meso-scale does not coincide with an additional pressure drop caused by friction between the polymer fluid and the reinforcement fibres.

In case the model is extended to a multi-scale model incl. macro-scale, however, an absorption of pressure by the thermoplastic fluid on macro-scale would directly cause a pressure decrease on meso-level, thus slow down impregnation and therefore provoke the creation of voids.

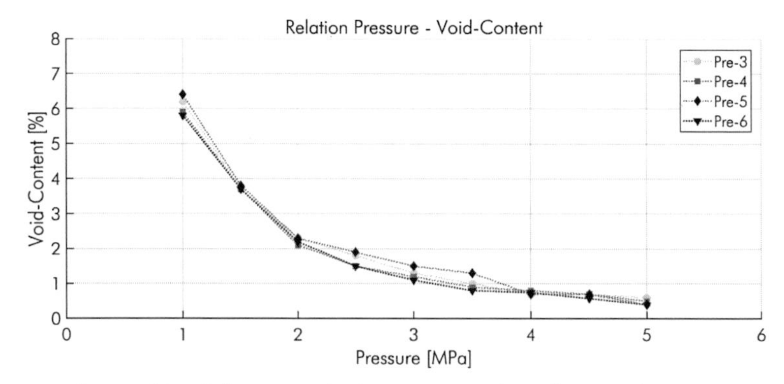

Figure 58 – Relation applied pressure and resulting void content

Process parameters – Influence of pressure on optimum fibre volume content

Concerning the influence of the applied pressure on the optimum fibre volume content, Figure 59 indicates an increase of the achievable mean fibre volume fraction with increasing pressure in form of a flattening function as already predicted in section 4.4. At the optimum fibre volume content in this context, the model predicts a completed impregnation throughout the whole ply thickness without excessive polymer fluid remaining.

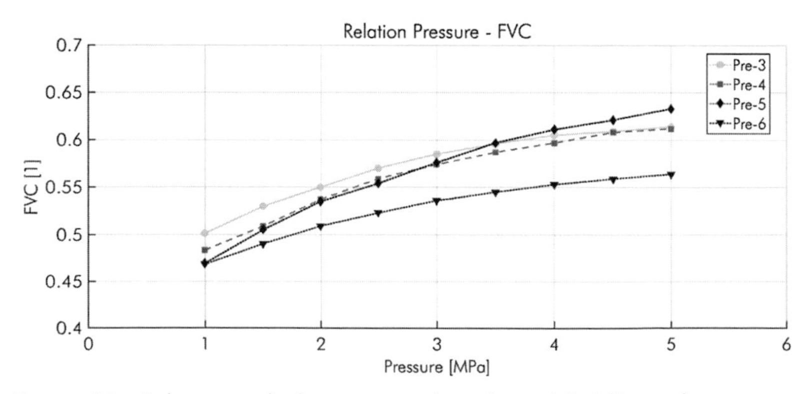

Figure 59 – Relation applied pressure and resulting global fibre volume content

When comparing the modelled data, Pre-6 stands out with lower optimum FVC values compared to the other textiles. Considering the compaction models in section 4.4, the relaxation curve for Pre-6 is noticeably flatter than for the other textiles. Here the explanation for the different fibre volume contents can be identified as well as the influence of the compaction behaviour.

Process parameters – Influence of temperature on time for impregnation

Beside pressure and time, the temperature and the directly linked viscosity take influence on the impregnation. To quantify the effect of viscosity, Figure 60 exemplifies the impact of a changing process temperature on impregnation time, which results in viscosities of 290 Pa s (at 390°C) respective 350 Pa s (at 370°C). The modelled durations for impregnation show an increase in dimensions between 7% and 12% at higher viscosities for pressures above 3 MPa. Compared to the applied pressure, the impact of the process temperature on the required impregnation time is slightly smaller, however, in an equivalent dimension.

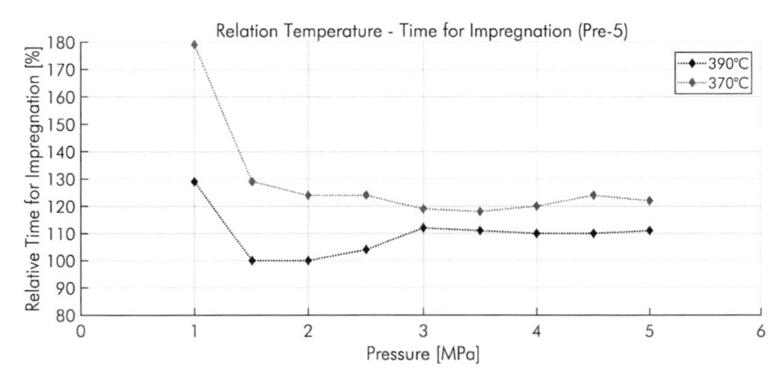

Figure 60 – Relation impregnation temperature and resulting time for impregnation

6.2.6 Aspects for model-validation

The previously discussed model addresses the impact of textile and process parameters on the process of impregnation and laminate quality. Due to the aforementioned simplifications, only the trends of the results can be considered for process development. For absolute values a model improvement is necessary. Therefore, a validation of the results based on absolute values is not applicable for most of the cases. In contrast, a trend-validation is recommendable, which has been performed in chapter 7 for some of the model results. A brief overview of procedures for model-validation is given below.

Two measurable aspects need to be considered for an examination of the influence of the applied pressure and the flow-path length on the progress of impregnation: (1) the resulting laminate quality that is measurable in terms of the void-content, which could be combined with a degree of homogeneity of matrix and reinforcement-fibres and (2) the impregnation-rate, respectively the time until impregnation is complete or stopped.

Concerning the resulting void-content, surely micrograph- or gravimetric-analyses provide adequate data to evaluate the influence of different pressures, as it was considered in session 7.2.2. An assessment of the degree of mixing certainly requires online image analysis of micrograph-sections, which has been done qualitatively in chapter 7.2.2, not quantitatively.

The determination of the required time for a completed impregnation at different pressures necessitates a suitable design of consolidation experiments to obtain reliable information. An alternative to investigate the progress of impregnation is the continuous acquisition of the laminate thickness during the impregnation process, as performed in chapter 8.1. This technique offers the advantage of an online-assessment and a fast validation, however with the drawback, that only a relative impregnation-progress is observable without absolute data about the void-content.

Model-validation concerning influence of impregnation length on required time for impregnation

In order to validate the modelled correlation between the flow path length and the required time for impregnation on meso-scale it is a challenge to identify the initial status of the textile arrangement, specifically the degree of mixing of thermoplastic- and reinforcement-fibres and the homogeneity of their distribution. This initial status should be considered, which presupposes the application of a two-dimensional model to also consider influencing flows in and around adjacent fibres and the different scales. Consequently, meso- and macro flow need to be studied in parallel, which is the reason that the validation of the modelled required times for impregnation with respect to the textile setup is not performed here. This kind of improved model could then be validated in consolidation experiments as performed in chapter 7, where void-contents of preforms with distinguishing flow path length are analysed after a specific impregnation time. For model-validation with explicit focus on the flow path length preforms containing equal fibres but with a different degree of spreading should be consolidated at equal process conditions. A different degree of spreading of the fibres can enable the variation of the flow path length without changing other textile parameters. It is recommended in this context as validation method.

Concerning the impregnation rate during the impregnation process the online acquisition of the laminate thickness in chapter 8.1 shows compaction with a progress between linear and quadratic behaviour during consolidation, which is in qualitative agreement with the modelled trend illustrated in Figure 56. This correlation should be focussed in further experiments.

Model-validation concerning influence of pressure on required time for impregnation

As discussed above, the influence of pressure on the required time for impregnation can be assessed by a subsequent void-analysis, which is performed for all laminates in chapter 7.2.2. The alternative method of an online thickness-acquisition as conducted in chapter 8.1 does not show a clear trend to increasing impregnation rates with increasing pressures which underlines the results of the impregnation model (Figure 57). Further experiments to evaluate the reliability of a correlation between the online thickness assessment and the required time for impregnation are recommended here.

The results of the void analyses show that high pressures are a prerequisite for a complete impregnation. Experiments in chapter 7.2.2 indicate that impregnation stops in textiles with long flow-path length, if the applied pressure is not above a certain level or if the absorption of pressure on macro-scale becomes too dominant. This complicates the investigation of the impact of the pressure on the impregnation rate and requires a higher number of samples.

Model-validation concerning influence of pressure on void content

For validating the model results about the influence of pressure on the resulting void content, experiments in chapter 7.2.2 are conducted including a subsequent void-analysis. Here, the applied pressure is varied between 1, 2 and 4 MPa. The assessment of the void-content delivers values comparable to the model-results for the textiles with small flow path length and with a fine and homogeneous distribution of matrix- and reinforcement fibres at high pressures of 4 MPa (Figure 70). Consequently, impregnation pressures for hybrid textiles shall be above 3-4 MPa as recommended by the model. For smaller pressures and for a low degree of homogeneity of the fibre distribution, a significant influence of the macro-flow was detected in the experiments. This again leads to the recommendation to implement a two-dimensional analysis of the impregnation, at least to model the process of impregnation for textiles with a low degree of homogeneity of the fibre distribution and for low applied pressures.

6.3 Solidification

This chapter describes the cooling step of the developed Isoforming process. As described in chapter 5, the cavity is transferred from the hot to the cold compression tooling which withdraws heat. This leads to a temperature drop inside the system, which again influences the properties of the polymer inside the composite. Cooling leads to shrinkage inside the matrix. Especially, matrix-rich regions can therefore suffer cracks or the evolution of voids. Furthermore, low cooling rates with associated growth of crystalline structures intensify this effect. The effects of cooling and the modelled temperature trajectory are described below.

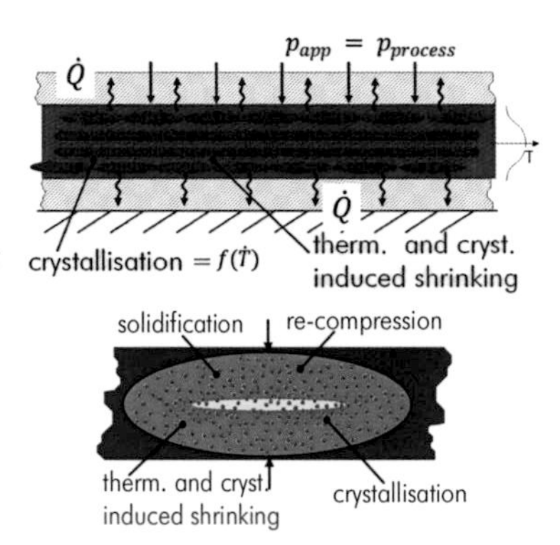

Figure 61 – Isoforming - cooling/ solidification

Modelling of temperature distribution during cooling

Modelling of the transient temperature distribution shall deliver process constraints regarding heating and cooling rates and not least the temperature distribution inside the composite laminate during cooling. The temperature distribution during cooling is of importance for an estimate of the creation of crystalline structures. Different cooling rates at the laminates top and bottom surface would lead to distortions, resulting from an uneven point of solidification and different degrees of crystallisation.

All parameters concerning the tooling were already shown in Table 22, so that only the composite parameters are changed to the laminate state as listed in Table 23. Figure 62 illustrates the temperature change inside the laminate during cooling, taking the evaluated thermal properties of the material into account. As for heating, the distribution of the temperature across the laminate thickness is homogeneous during cooling. The high thermal mass of the compression toolings compared to the cavity only leads to a slight temperature increase for them.

Table 23 – Parameters for transient thermal simulation of cooling step

			Source
Composite (consolidated state)			
Initial temperature (151G / AE250™)	[°C]	400 / 360	
Thermal diffusivity	[mm²/s]	$a(\varphi, T)$	Figure 26
Fibre volume content	[1]	0,6	
Cavity (steel 1.2312)			
Initial temperature	[°C]	400 / 360	
Tooling components (steel 1.2312)			
Initial temperature	[°C]	100	

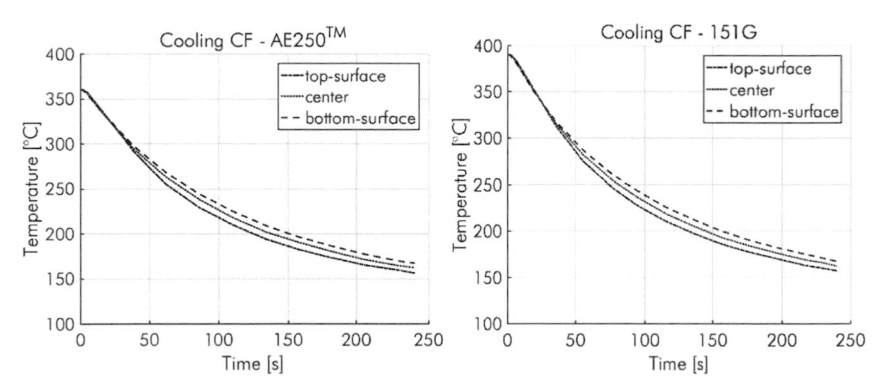

Figure 62 – Temperature distribution of laminate during cooling

The figures demonstrate that demoulding is possible after 4 min of cooling, when the temperature falls below the glass transition region. The limiting factor for the time required for cooling are the dimensions of the cavity. Reducing the temperature of the compression tooling only contributes to a small extent to higher cooling rates. With a tooling temperature of the cold tooling of 80°C, the resulting temperatures inside the laminate after 4 min are only 5 K smaller than for a tooling temperature of 130°C. Comparable data are obtained for both PEEK polymers. For 151G, which has an initial process temperature of 390°C, the final temperature after 4 min is only 5 K higher than for AE250™-polymer, being processed at 360°C.

Beside productivity and the resulting demand for high cooling rates, a counteracting requirement usually is a high degree of crystallinity, resulting from low cooling rates. This area of conflict needs to be analysed. The modelled time-dependent cooling rates for discrete time steps are displayed in Figure 63.

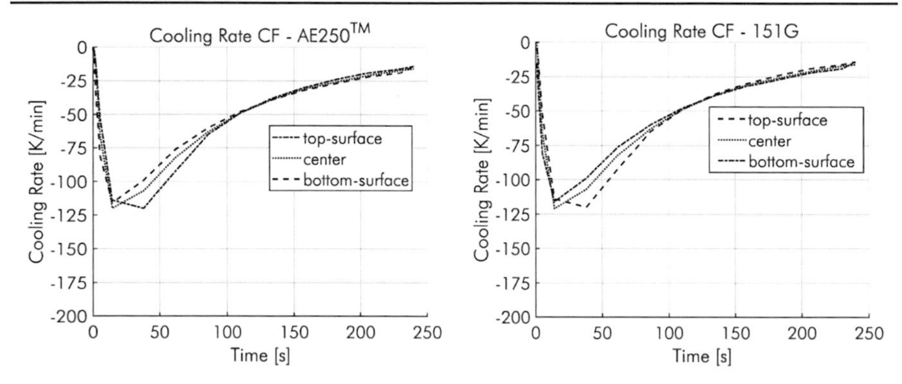

Figure 63 – Modelled cooling rate in laminate

The resulting cooling rates can be further used to estimate the local degree of crystallinity. Section 3.2 delivers crystallisation characteristics, which allow an approximated forecast of 25 J/g for the AE250™-polymer respective 37 J/g for 151G. Taking 130 J/g as reference value for the total heat of fusion, the resulting degree of crystallinity are presumably in regions of 19% respective 28%. Both values are conservative approximations since they apply for the raw polymers without the possible contribution of filaments to crystallisation.

6.4 Conclusion – model for consolidation of hybrid textiles

Concluding, the results of the consolidation model are summarised.

Heating

Concerning the modelled transient temperature distribution inside tooling and preform, the obtained data show that the dimensioning factor for the heating rate is the tooling. Heat transfer inside the preform is faster than heating the cavity. Regarding the heating trajectory, it is recommendable to wait about one minute until a homogeneous temperature field is achieved, if the process temperature is attained at the surfaces of the preform. For the required durations for heating and cooling, an AE250™-polymer laminate does not provide a significant process acceleration despite the smaller processing temperatures. A reason for this is the smaller temperature difference between laminate and tooling in the final stage of the heating respective cooling process, which leads to comparable heating or cooling rates compared to laminates with classic PEEK and processing temperatures up to 400°C.

Impregnation

The impregnation model delivers reasonable results for the void content, the fibre volume content and the relative impregnation time, despite the mentioned simplifications in the sub-models (permeability, compression behaviour and flow path length). It shows that the most effective method to reduce impregnation times is the reduction of flow path lengths and thus an increase of the degree of pre-mixing of the hybrid textiles. Increasing the pressure or reducing the viscosity by higher temperatures provides a comparable impact concerning impregnation time. A temperature increase of 20 K has shown to reduce the required impregnation time by 7-12%, which is comparable to the effect of doubling the pressure from 2 to 4 MPa. However, for both effects, their impact is significantly smaller than for adjusting the flow path length.

Beside the reduction of uncertainties of the sub-models, the consideration of a multi-scale arrangement of carbon-fibres and thermoplastic fluid, can increase the quality of the model. Especially on macro-scales the pressure absorption of the fluid shall be considered for model improvement. Generally, a differentiation between macro- and meso-scales is recommendable.

Cooling

For the cooling phase, the modelled data show that a temperature decrease until glass transition temperature is feasible in 3 to 4 min. The cooling rate has its maximum value in the beginning of the cooling phase. Subsequently, the rate decreases successively. Depending on the necessary degree of crystallinity, also smaller cooling rates are possible to support the creation of crystalline growth.

7 Sensitivity analysis of consolidation parameters

Beside a model-based approach and the development of the process itself, this thesis also comprises an experimental sensitivity analysis about the effects of the process parameters pressure, impregnation time, temperature and material properties on the void content and resulting mechanical properties. The following chapter consequently includes the description of the Isoforming process for consolidation experiments with the hybrid textile preforms Pre-3, Pre-4, Pre-5 and Pre-6. By means of these experiments, it is also possible to validate the developed impregnation and thermal models.

Experiment planning and execution

For process analysis, all four introduced textiles are consolidated by the developed Isoforming process. To evaluate the influence of the different parameters, impregnation time, applied pressure, temperature and preform thickness, the test matrix shown in Table 24 is considered. All laminates are manufactured as biaxial laminates with fibre orientations in ±45° direction, except for Pre-6 which contains four fibre orientations in a single pre-cut layer.

For executing the thermoforming process, release agent was applied to the cavity surface before the preform was inserted including ejector and pressure boards. Placing the cavity between the compression toolings was performed manually. Also, the transfer between hot and cold compression toolings was done hand operated. During the heating phase, the pressure was kept to a minimum. Impregnation pressure first was applied, when it was assured that all thermoplastic filaments are molten. During the impregnation stage, temperature was kept constant. The subsequent transfer stage then took approximately 30 s, which did not result in a decreasing temperature in the cavity. From the beginning of the cooling phase in the cold compression tooling, a temperature drop was observed as predicted.

Table 24 – Process parameters

Material	Process Conditions Δt [min]	p [MPa]	T [°C]	Stacking
Pre-3	10	1; 2; 4	400	$(+45/-45)_{2s}$
	20	1; 2; 4		
Pre-4	5	4	400	$(+45/-45)_{2s}$
	10	1; 2; 4		
	20	1; 2; 4		
Pre-5	5	2; 4	400	$(+45/-45)_{4s}$
	10	1; 2; 4		
	20	1; 2; 4		
	10	1; 2; 4	380	
Pre-6	5	1; 2; 4	400	$(90/+45/0/-45)_{s}$
	10	1; 2; 4		
	20	1; 2; 4		

An exemplarily illustration of the pressure trajectory and the tooling temperature 4 mm below the preform surface is shown in Figure 64. In this example, a temperature of 380°C was reached after 8 min as modelled in chapter 6. With increasing time, temperature increased until 390°C. Concerning the cooling rate, temperature gradients of -180 K/min in the beginning of cooling and -14 K/min after 3 minutes of cooling were observed, which is in good agreement with the modelled values (see chapter 8.2).

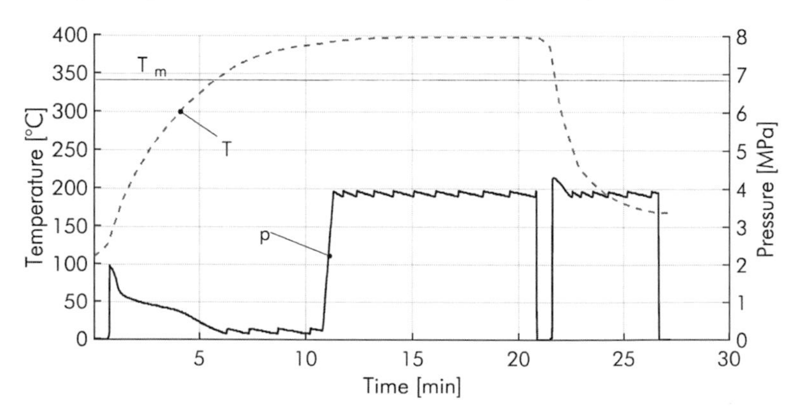

Figure 64 – Exemplary process diagram during thermoforming (impr. time 10 min, pressure 4 MPa, hot compr. tooling temp. 400°C, cold compr. tooling temp. 140°C)

Analysis

For analysing the effects of process parameters and material influence, the thickness of all laminates was measured at four points. Furthermore, specimens were extracted for measuring the local fibre volume content, for preparing polished micrograph sections and for measuring the apparent inter-laminar shear strength (ILSS). An illustration of all significant positions of the composite laminates is given in Figure 65.

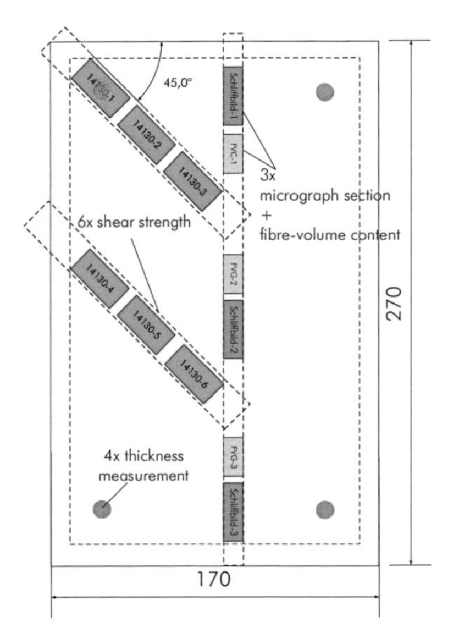

Figure 65 – Analysed sections of manufactured laminates

7.1 Influence of textile configuration

All four textile configurations are analysed in terms of their microstructure and their resulting laminate thickness in this chapter.

Table 25 provides examples of the polished cross-sections of laminates, which were manufactured at 4 MPa with impregnation times of 10 and 20 min. Beginning with Pre-3, an inhomogeneous impregnation with several voids and matrix-rich regions is observed, which improves with increasing time. A likely explanation is the inhomogeneous mixing of matrix and carbon fibres in the preform. Due to fluctuations in the impregnation distances (see section 4.2), the polymer could not penetrate all rovings and it accumulated in several regions.

Table 25 – Polished cross-sections of laminates (4 MPa) from different hybrid textiles

	Pre-3	Pre-4	Pre-5	Pre-6
10 min	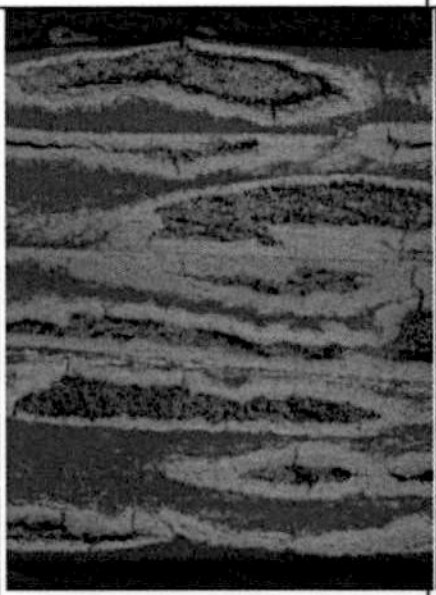		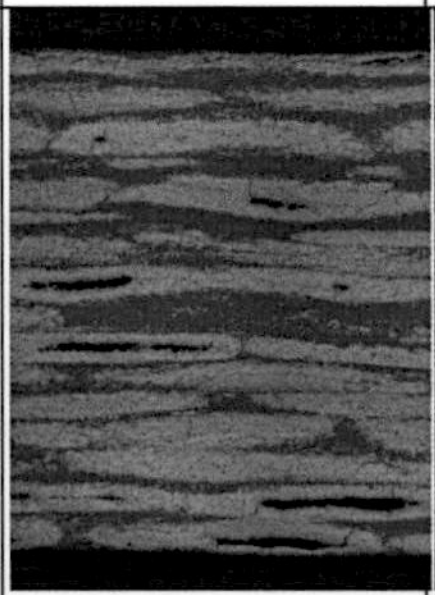	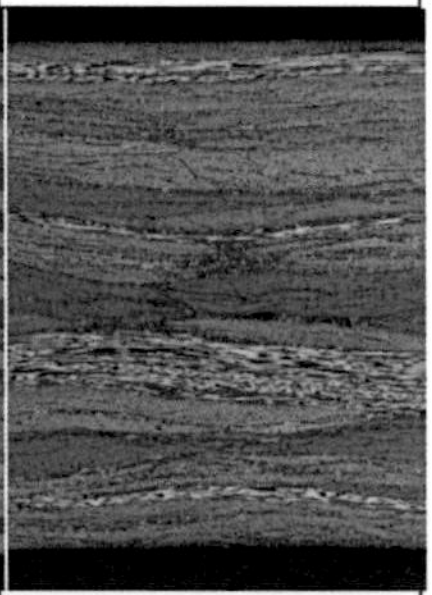
20 min	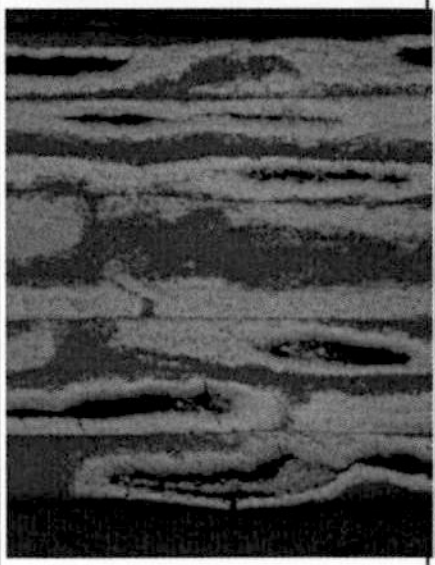			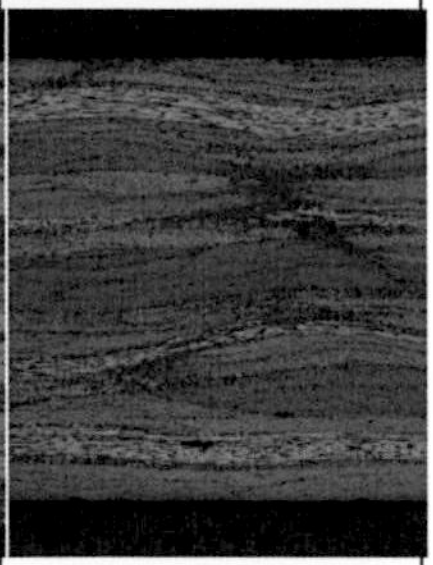

Regarding Pre-4, a relatively homogeneous fibre–matrix distribution is achieved. Longer durations for impregnation lead to an improvement. The void-free microstructure underlines the good performance of the material in terms of homogeneity.

Pre-5 contains the same raw materials as Pre-3. Thus, a different impregnation behaviour is directly linked to the textile structure, which is more consistent than for Pre-3 as shown in section 4.2. Here a considerably higher degree of homogeneity is observable. Rovings are spread relatively equal. Matrix accumulations or unimpregnated rovings also remain here, however to a smaller extent than for Pre-3. The micrograph-section shows that even thick 12k carbon fibres can be impregnated, if the degree of spreading is high enough and if the matrix fibres are distributed equally on top of the carbon fibre rovings.

The most regular arrangement of filaments and matrix is delivered by Pre-6. This material enables an almost void-free impregnation without any matrix accumulations. Even single rovings are difficult to identify, due to the good distribution of filaments and matrix.

Regarding the resulting fibre volume content, no significant effect of the process parameters could be determined. Only the material arrangement, respectively the amount of the integrated fibres, had an impact on the resulting fibre volume content. In this context, the side-by-side arrangement in Pre-3 and Pre-5 resulted in a fibre volume fraction of 53%, while the hybrid yarn textiles in Pre-4 and Pre-6 lead to 45% and 56%, respectively. The relatively small fraction of Pre-4 potentially simplifies the impregnation process, although this has a negative impact on the lightweight aspect. Comparing the results for the hybrid yarn textiles with the assumed fractions from the yarns in Table 12 shows that, especially for Pre-3 and Pre-6, values between pretended and measured values deviate. While for Pre-3, the measured fibre volume content is 7% lower than pretended, which would benefit impregnation, the measured fibre volume fraction of Pre-6 is 7% higher than pretended. This normally decelerates the impregnation process. The fluctuations of the fibre volume fractions presumably result from local fluctuations of the titre of the thermoplastic yarn, which again leads to a demand of an intake control of the hybrid yarns prior processing. Concerning the impregnation model in chapter 6, these deviations do not affect the modelled results, since the modelled local fibre volume contents and all resulting parameters are based on the compression–strain curves, which do not consider the global fibre volume content as a direct input parameter.

Concerning the fluctuation of thickness, all laminate thicknesses vary in dimensions between 4% and 9% in common, which is represented in Figure 66. Reasons for fluctuations can be squeeze flow of the matrix and fibre bundles in in-plane direction or an unbalanced press respective tooling with small deviations from perfect horizontal alignment. It is remarkable that for Pre-3, the thickness shows higher deviations than for laminates from other textiles. This again illustrates the significant influence of the homogeneity of filament distribution in the hybrid textile on resulting laminate properties.

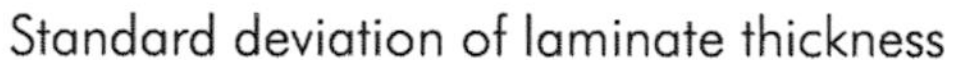

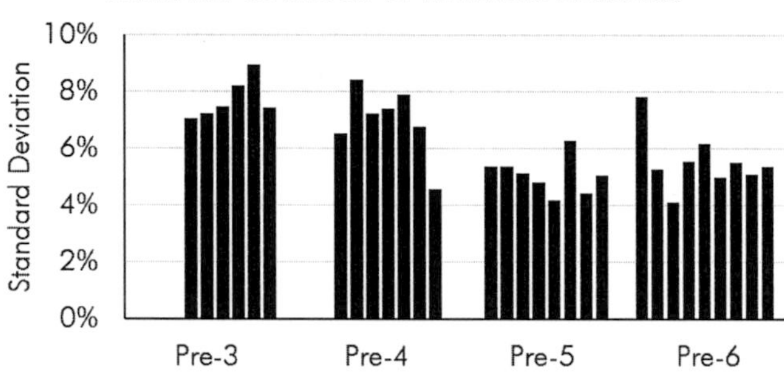

Figure 66 – Standard deviations of laminate thicknesses

Conclusion

Concluding, the arrangement of reinforcement and matrix fibres inside the preform has a dominant influence on the laminate properties. Inhomogeneous distributions of reinforcement and matrix fibres result in varying flow path lengths and resulting matrix accumulations or unimpregnated regions. Consequently, prerequisites to maintain high-quality laminates are, first, to minimise flow path distances by a high degree of pre-mixing and, secondly, a high homogeneity of the distribution of both fibre types.

7.2 Influence of process parameters

The following sub-section analyses the influence of both process parameters: impregnation time and applied pressure on the resulting laminate. Their impact is studied based on polished micro-sections, the measured laminate thickness and the resulting void content.

7.2.1 Influence of impregnation time

Influence on laminate quality

Having a closer look at the macro-structure and meso-structure of the laminates illustrates the impact of the duration of impregnation. Table 26 displays polished cross-sections of all four material specifications, which were processed at 4 MPa with different impregnation durations (5, 10 and 20 min).

Since Pre-3 and Pre-5 contain the same raw materials, only their textile configurations lead to differences in their structure. The textile of Pre-5 shows significant improvement in terms of homogeneity with increasing time compared to Pre-3. After 20 min, only very few rovings contain voids, while in Pre-3 still almost each roving contains unimpregnated areas even with increasing impregnation time. Moreover, still large regions of matrix accumulations can be observed here, which supports the assumption of a pressure absorption of the fluid on macro-scale and thus a reduced pressure acting inside the roving. A higher degree of mixing or a better stacking in out-of-plane direction of thermoplastic and carbon-fibres should be considered for improvement.

Pre-4 with its formerly pre-mixed hybrid yarns also shows an improved homogeneity. After 10 min of impregnation, only a small number of voids remain. Ten more minutes for impregnation lead to a void-free laminate. Accumulations of the polymer still remain, however with a remarkably smaller amount, than for the arrangement in Pre-3.

Pre-6 shows an excellently homogeneous micro-structure, with only few voids already after 5 min. After 10 min, most voids vanished. Together with Pre-5, its high degree of pre-mixing enables the shortest required impregnation times of all textiles. Regarding the homogeneity of the fibre–matrix distribution, Pre-6 provides the highest quality of all textiles.

Table 26– Polished cross-sections after different impregnation times at 4 MPa

	5 min	10 min	20 min
Pre-3	/	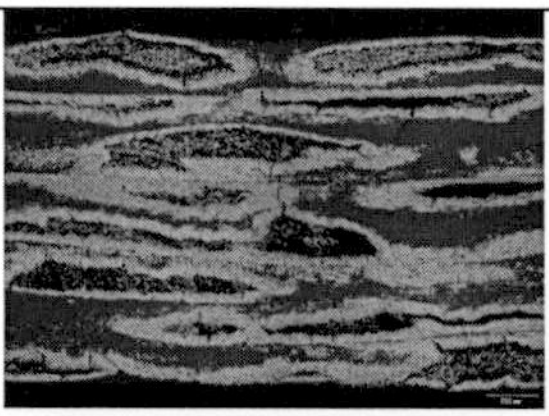	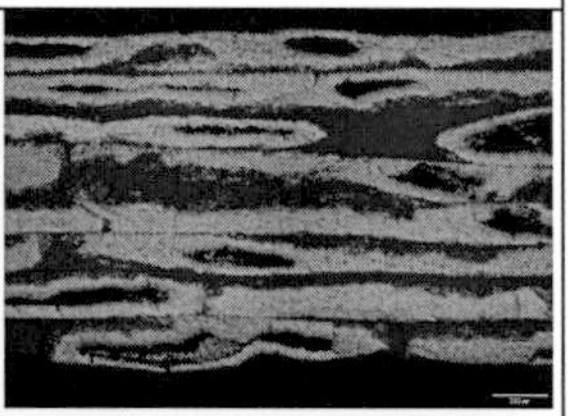
Pre-4	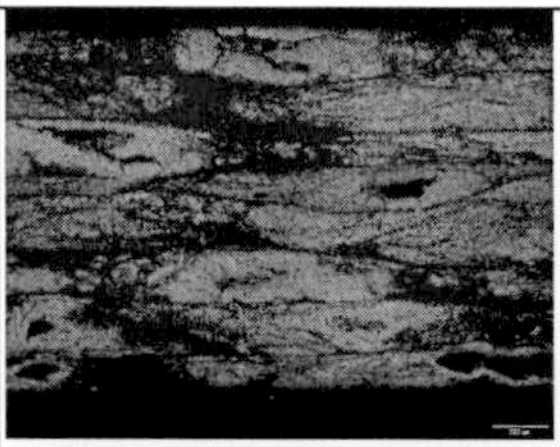	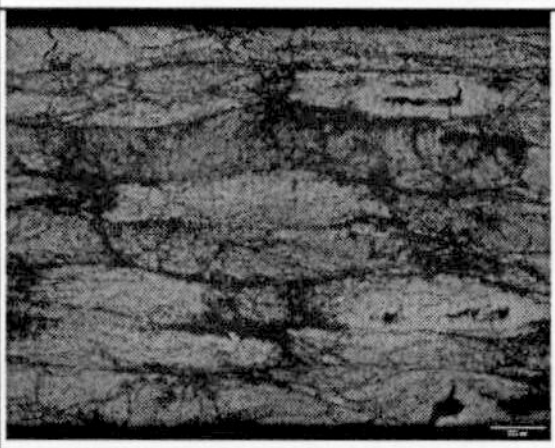	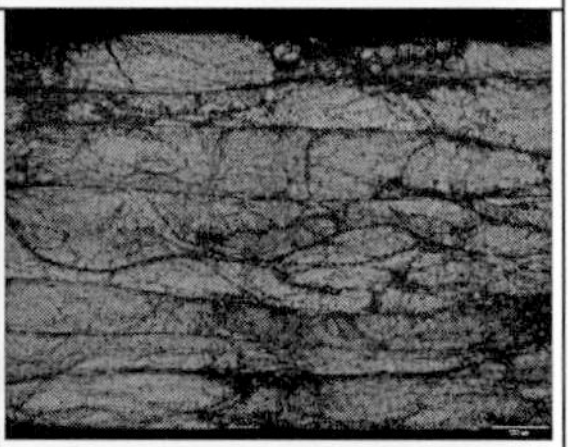
Pre-5		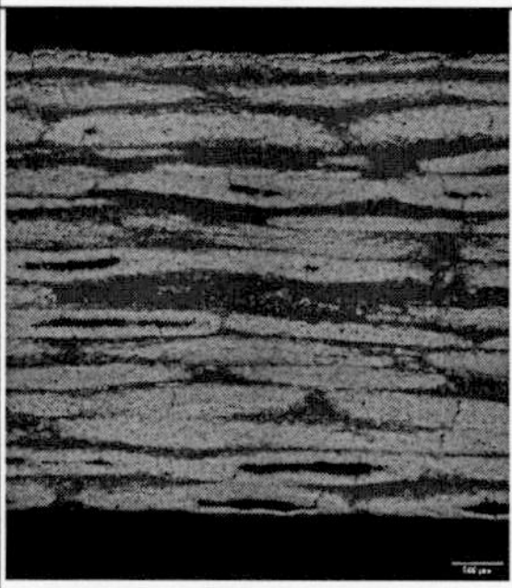	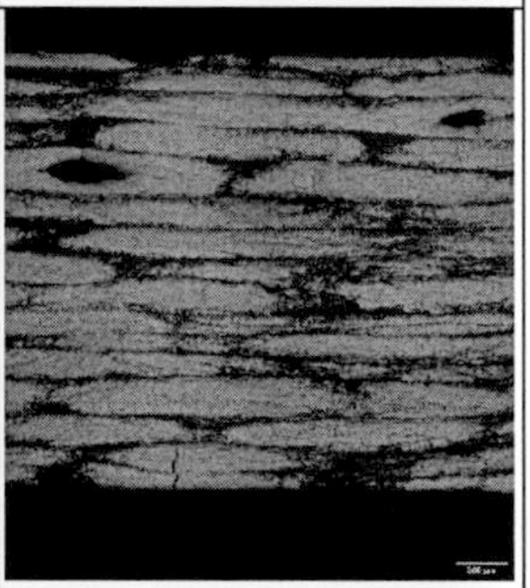
Pre-6	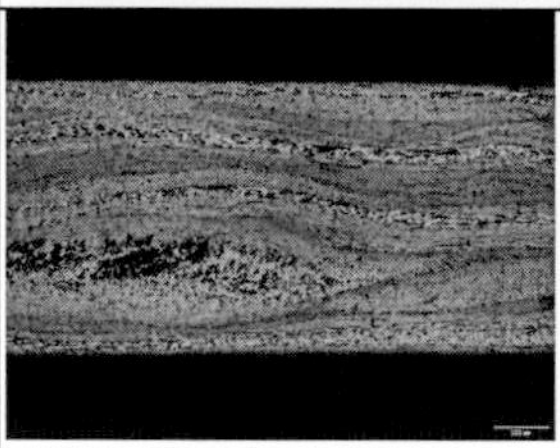	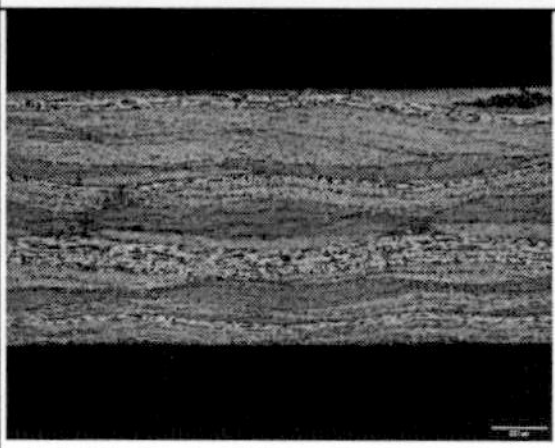	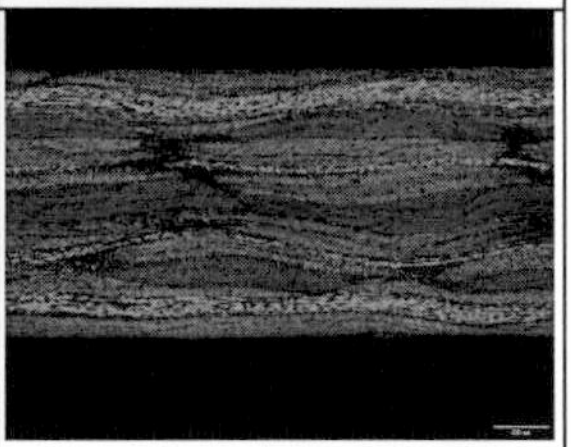

Influence on void content

The measurement of the void content was performed by picture analysis of polished micrograph-sections. Three specimens per laminate were considered per measured void content. Their mean values are represented in Figure 67 (see Figure 65 for the locations of the extracted specimens).

The side-by-side arrangement in Pre-3 and Pre-5 places the condition of long impregnation distance. The data point out that voids can only be reduced by long impregnation durations in combination with high pressure. Concerning the duration of impregnation, the obtained data show that impregnation times below 10 min lead to unacceptably high void contents. Especially, the Pre-3 requires at least 20 min for impregnation.

Both textiles with pre-mixed hybrid yarns provide low void contents below 3% already after 5 min (Pre-4) and 10 min (Pre-6).

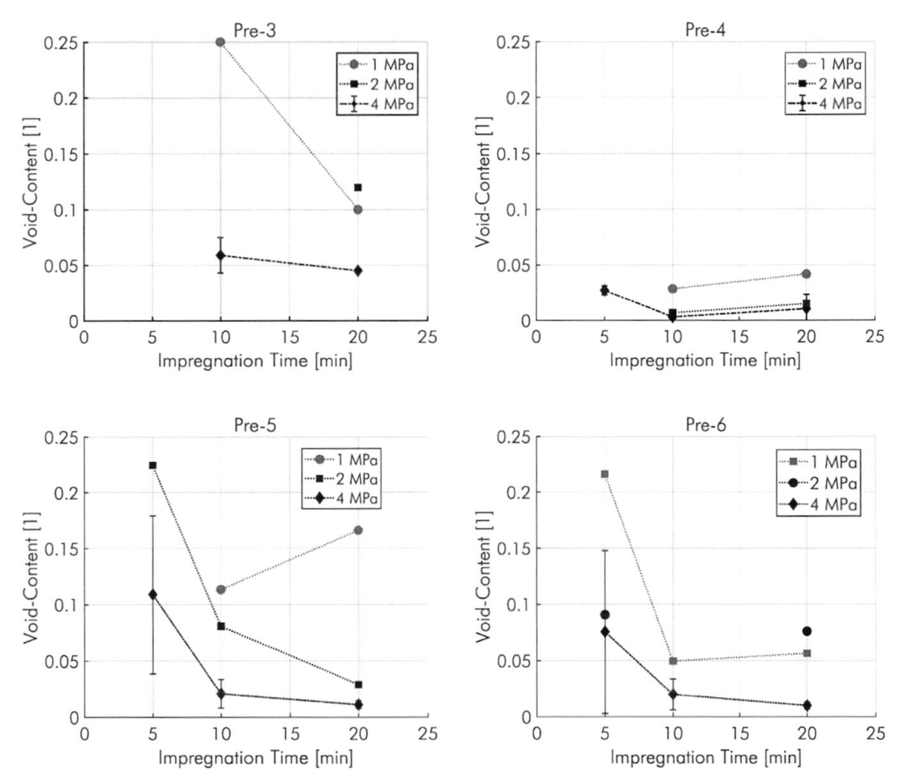

Figure 67 – Void content with respect to impregnation time at different applied pressures

A general conclusion is that if the applied pressure is low, voids are not compressed any further even after longer impregnation times. Preforms with long impregnation length benefit from longer impregnation times, especially for lower pressure (1-2 MPa). The level of porosity however remains too high in this case. Preforms with short impregnation lengths, on the other hand, obtain low void contents already after short time (see Figure 68).

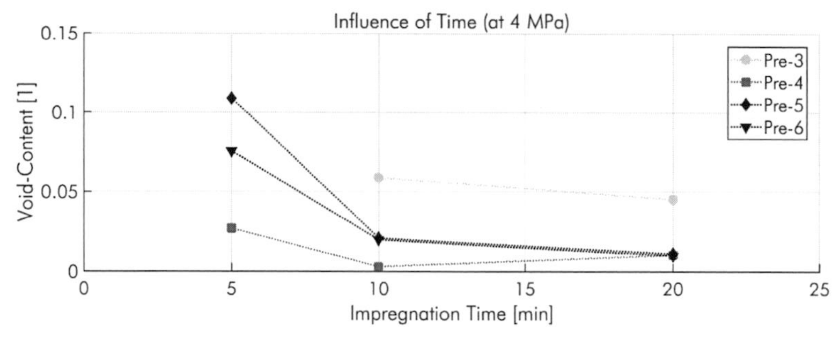

Figure 68 – Void content in relation to impregnation time at 4 MPa

Influence on laminate thickness

Concerning the thickness of the manufactured laminates, no significant effect of the impregnation time could be detected. As it is exemplified in Figure 69, all process configurations deliver comparable laminate thicknesses, which only vary in dimensions of the standard deviation.

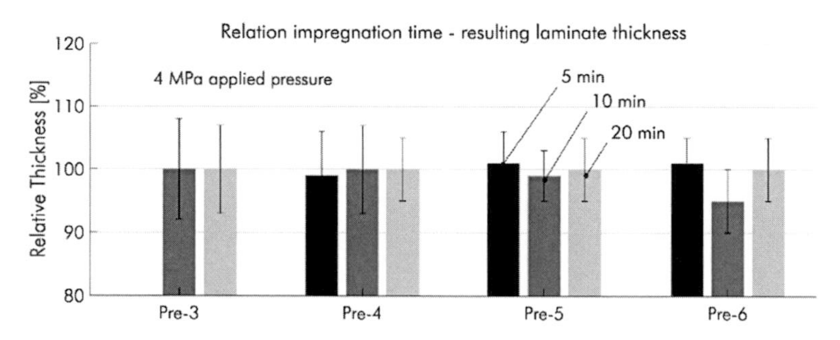

Figure 69 – Relation between impregnation time and resulting laminate thickness

7.2.2 Influence of pressure

Influence on laminate quality

Table 27 gives an overview of the laminate cross-sections at varying applied pressures after 10 min of impregnation. Here, impregnation is not complete for Pre-3 to Pre-5, so the influence of the acting pressure is possible to outline. Beside Pre-6, which delivers an almost equal quality despite changing pressures, an increasing pressure has a significant impact on the other laminate's qualities. For Pre-3 and Pre-5, especially the increase from 1 to 2 MPa has a remarkable effect, resulting in a higher degree of impregnation.

With growing pressure to 4 MPa, particularly Pre-5 benefits in terms of impregnation quality. The less homogeneous Pre-3 benefits less from the higher applied pressure.

Table 27 – Polished cross-sections at different pressures after 10 min

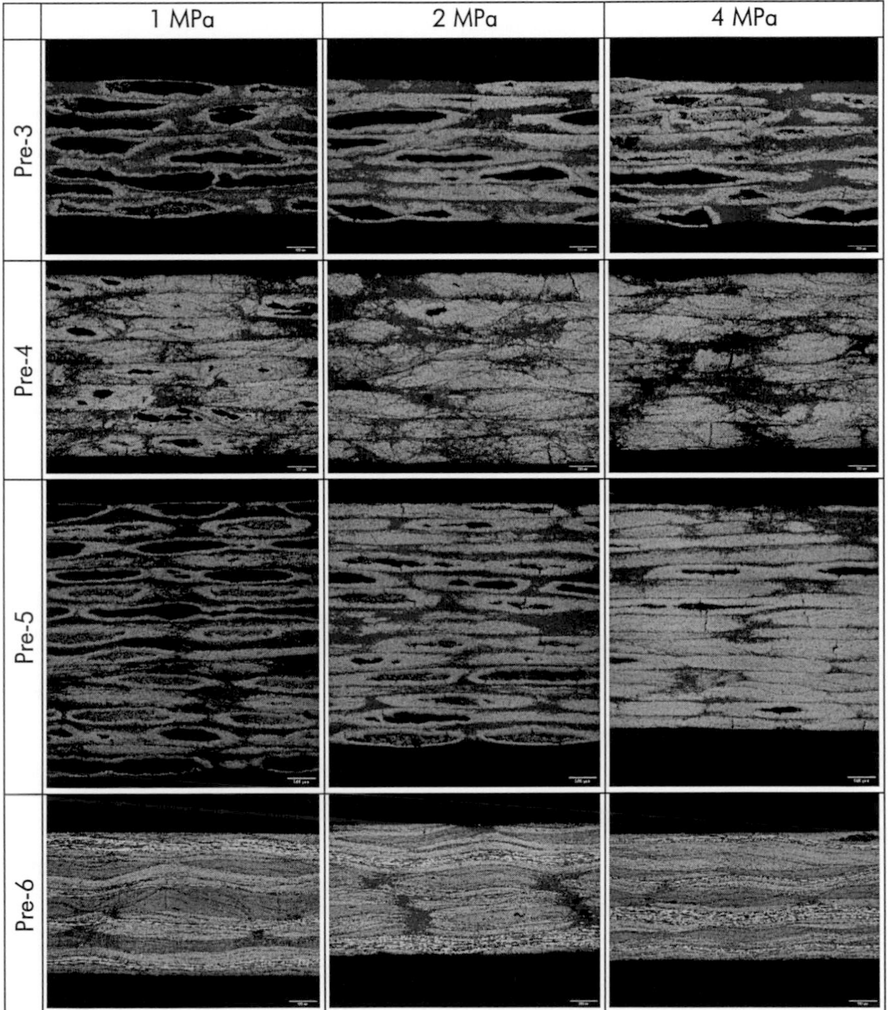

Influence on void content

Looking at the measured values of the void contents from the polished cross-sections shows a significant impact of increasing pressure on porosity, as illustrated in Figure 70. Both textiles with long impregnation distance (Pre-3 and Pre-5) benefit remarkably from an increasing pressure in terms of the void content. Doubling the pressure from 2 to

4 MPa enables a reduction of the void content from 10% to 6%, respectively, from 8% to 2% after 10 min of impregnation. Also, the textiles with pre-mixed hybrid yarn (Pre-4 and Pre-6) benefit from the pressure increase. A doubled pressure to 4 MPa enables impregnation already after 10 min, due to void contents below 3%.

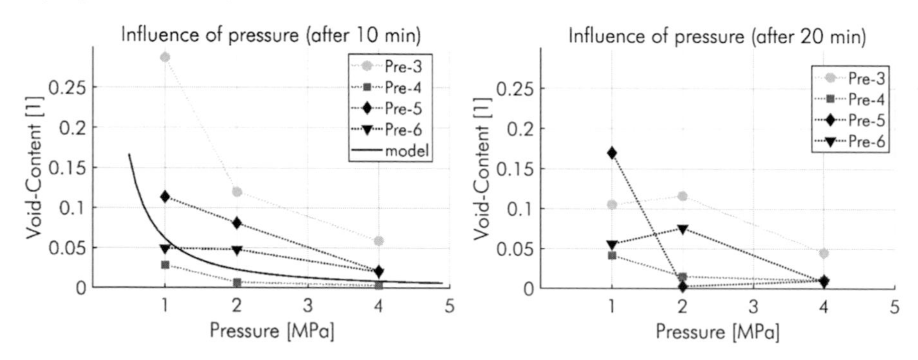

Figure 70 – Void content with respect to applied pressure

The experiments show that pressures below 2 MPa are not sufficient to create a void-free laminate for none of the four textile configurations. This correlates with the evaluated relations between applied pressure and resulting fibre volume content in section 4.4, where a pressure of 1 MPa was not sufficient to gain fibre volume contents above 50%. Pressures of 4 MPa and presumably higher enable a significant reduction of voids, which sustains the impregnation model (Figure 70 left), including the assumption of a compression of voids with increasing applied pressure.

Influence on laminate thickness

In terms of thickness, Figure 71 demonstrates the relation between the applied pressure during the impregnation stage and the resulting laminate thickness. A clear trend to higher degrees of compaction at higher pressures can be observed. So, the applied pressure has a higher influence on the final thickness respective compaction compared to the impregnation duration described in the previous chapter.

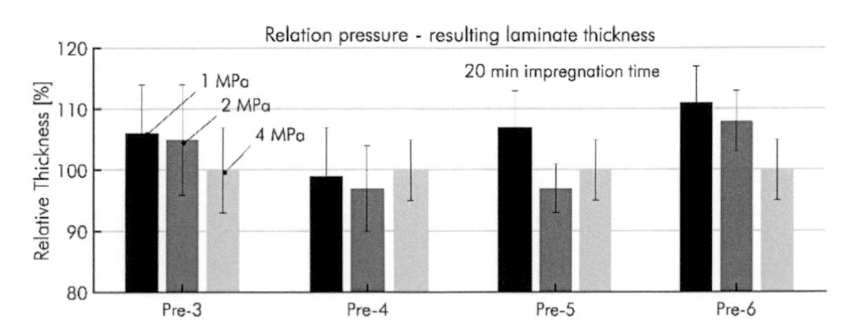

Figure 71 – Relation between applied pressure and resulting laminate thickness

7.2.3 Influence of solidification on internal stresses

When analysing the micrographs, it is conspicuous that some of the laminates have micro cracks within some rovings. High-performance thermoplastics, as PEEK or PEKK, require high processing temperatures and cross a high temperature range from solidifying during consolidation until room temperature. This large temperature range can result in higher internal stresses than for thermoplastics with lower processing temperature. Cracks on micro-scale can appear induced by the process of solidification due to the different coefficients of thermal elongation along and orthogonal to the fibre orientation. Taking the results of sections 3.5 and 3.6 into account, the product of CTE and Young's modulus of PEEK increases significantly when the temperature falls below 220°C during consolidation and even more below the glass transition region.

In the experiments, the cracks are detected exclusively in laminates containing 12k rovings, which leads to the assumption that the roving size or its degree of spreading, respectively, its degree of homogeneity, has an effect on the cracks. Looking on macro-scale, adjacent plies in Pre-3, Pre-4 and Pre-5 have an offset of 90° in fibre orientation. Consequently, those adjacent plies counteract the process-induced shrinkage of a single ply. Shrinking orthogonally to the fibre orientation is supressed. This leads to residual stresses which increase with higher stiffness of individual plies. For Pre-5 and especially Pre-3, the single-ply thickness presumably became too high, resulting in a too high stiffness and consequently appearing micro-cracks. The improved homogeneity in Pre-5 and especially Pre-4 and Pre-6 lead to less or even no micro-cracks, which underlines this hypothesis. Also, the lower single-ply thickness probably contributes to a crack-free laminate.

Table 28 – Appearance of micro-cracks in polished cross-sections

	Side-by-Side with 12k-Rovings		Commingled Yarns
Pre-3	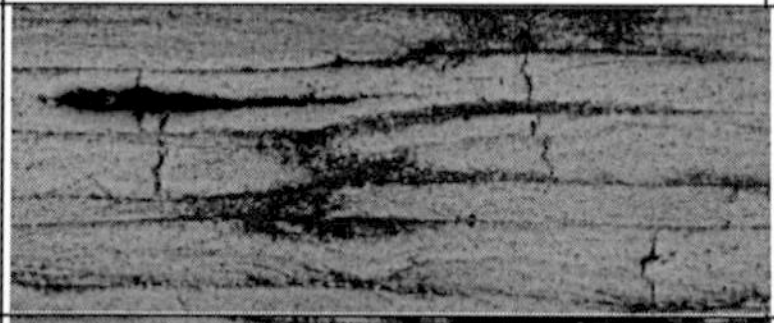	Pre-4	
Pre-5		Pre-6	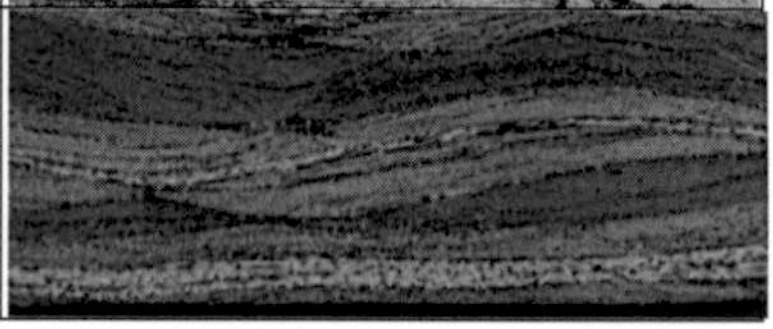

Therefore, the side-by-side arrangement with 12k fibres can only be recommended, if adequate and reproducible spreading can be assured. Further improvement can be gained by thermoplastic matrix filaments with smaller diameter. Alternatively, a pre-mixing of reinforcement and thermoplastic fibres is recommendable. If this is not taken into account, the probability of micro-cracks increases significantly.

7.3 Influence of the void content

The void content and the fibre volume content are values, which illustrate a laminate's quality by discrete values. However, interpretation of these values concerning their influence on the mechanical properties is a challenge. The diagrams below illustrate the void content with respect to the resulting mechanical properties in terms of the apparent inter-laminar shear strength (ILSS). The mechanical tests were realised by short-beam tests according to DIN EN ISO 14130, which is a three-point bending test with reduced beam length to cause failure by inter-laminar shear. This experimental setup derives a comparable value, which allows conclusions about the influence of voids.

The specimens from different textiles were not comparable to each other. They distinguished, for example, in terms of fibre orientation or the number of single plies. The specimen thickness governs the required distance between the supports for the short-beam test. Specimens taken from Pre-3 to Pre-5 all had a fibre orientation of 0/90 and a symmetric lay-up in common. Pre-6 contains fibre orientations in 0/45/90/-45 degree. Pre-5 and Pre-6 are composed of 16 plies, while Pre-3 and Pre-4 are made from 8 plies. All specimens had a symmetric lay-up in common. Results of the short-beam tests are shown by Figure 72. Each dot represents the mean value from six specimens.

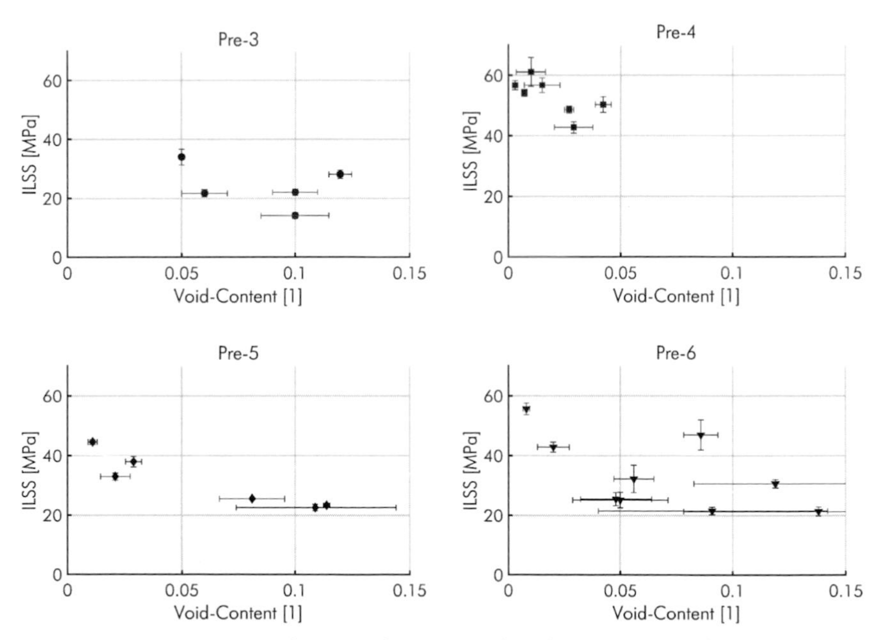

Figure 72 – Inter-laminar shear strength with respect to void content

The results indicate a close dependence between the void content and the resulting ILSS. As mentioned above, the values of the ILSS of the different textile materials are not entirely comparable to each other. However, for all configurations, a clear increase of ILSS is observable with decreasing void content.

In this context, Pre-3 and Pre-5 show a comparable behaviour, which makes sense, since both consist of the same material configuration and only distinguish in terms of their textile configuration. This supports the hypothesis of a direct correlation between void content and ILSS respective mechanical properties. Compared to Pre-4 and Pre-6, the smaller absolute values can result from matrix accumulations or micro cracks.

For Pre-4, the obtained ILSS are almost equal for different process configurations, which resulted in void contents below 2–3%. This supports the recommendation of reducing void contents below this threshold. Also, for Pre-6, the values with the smallest void content show the highest inter-laminar shear strength.

Regarding void contents above 5%, the values for ILSS remain on a low level roughly unaffected by a further increase of the void content.

Illustrating all measured data in Figure 73 again underlines the correlation between ILSS and void content. For modelling, an exponential behaviour delivers the best description of this relation.

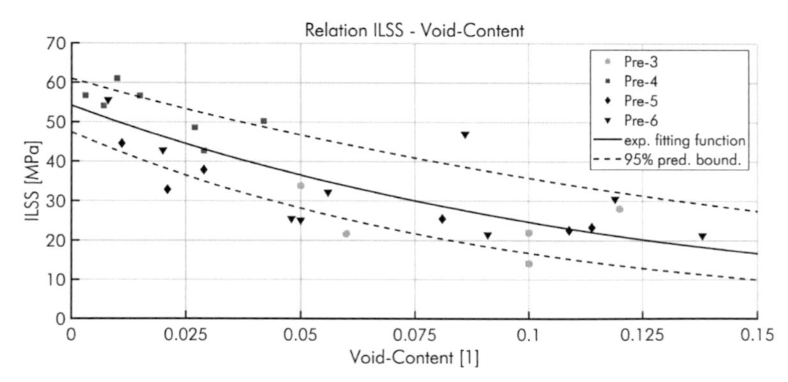

Figure 73 – ILSS with respect to the void content including fitting function

7.4 Conclusion – sensitivity analysis

Influence of material properties

In order to draw a conclusion, the experiments show that the textile properties have significant influence on the laminate properties. Parallel to the flow path length, the homogeneity of the thermoplastic and carbon fibre distribution has a significant impact on the resulting laminate quality. Beside voids, cracks provoked by internal stresses during the cooling cycle need to be prevented. The results indicate that accumulations of the polymer need to be minimised and ply thicknesses must not exceed certain thicknesses to avoid local crack generation. These threshold ply thicknesses and accumulation limits need to be taken into account in the future.

Influence of process parameters

The results further show that even preforms with small flow path length require minimum 10 min for impregnation, if a homogeneous distribution without voids shall be produced. For improvement of the laminate quality, an increase in applied pressure showed a higher impact than the extension of impregnation time. Generally, the results lead to the suggestion to apply high pressures of 4 MPa or even above, which coincides with the results of the models in section 6.2.4.

Influence of void content on mechanical properties

Finally, the investigation about the influence of the void content on the resulting mechanical properties indicates that a porosity below 2% shall be targeted to achieve the highest mechanical integrity of the laminate. A void content in ranges between 2% and 3% can already lead to a reduced ILSS in dimensions of 10–20%.

8 Process analysis

In this chapter, different approaches for analysing phenomena occurring during consolidation are investigated. The resulting detections act as validation for the developed consolidation model, for the potential improvement of the model in the future and for the general understanding of the manufacturing process and its improvement.

Tracking the thickness of the laminate during impregnation allows drawing conclusions about the degree of impregnation. Furthermore, a local determination of the aggregate state of the polymer shall help to verify a thermal analysis for heating the preform and cooling the consolidated laminate. This will allow approximations of the required time to heat and to cool, as well as to estimate the resulting local cooling rate, which influences the degree of crystallinity. Another objective of the process analysis is to examine the interactions between fibres, polymer fluid and entrapped air, which is done by a real-time visual inspection.

8.1 Online thickness acquisition

The hypothesis that tracking the thickness of the stacking during impregnation allows drawing conclusions about the degree of impregnation is examined in this chapter. The following experiments show a link between the laminate thickness and the penetration of the flow front.

Experiments

In order to investigate the correlation of the laminate thickness and the progress of impregnation, temporal thickness changes were tracked during the thermoforming cycles in chapter 7. For this, an eddy current sensor eddyNCDT3005 by micro-epsilon was attached to the Isoforming tooling, being able to measure dimensional changes with a resolution of 15 μm. Since a central positioning of this sensor at the laminate area is not possible, it is placed close to guiding pillars of the upper compression tooling. This of course results in uncertainties of the absolute measured thickness, since deviations from an entire horizontal arrangement of the upper tooling or the press would lead to amplified errors of the absolute measured values. Therefore, the data should only be considered as relative values. During each laminate manufacture, the current thickness was measured.

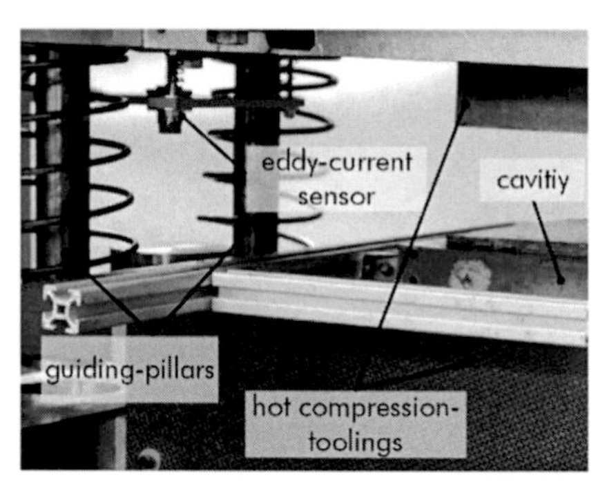

Figure 74 – Position of eddy-current sensor in Isoforming-tooling

Results and interpretation

Figure 75 illustrates the thickness change for all four textile configurations at process temperatures of 390°C, 4 MPa applied pressure and a duration for impregnation of 20 min. The characteristic of the thickness change in Figure 75 is qualitatively representative for each illustrated textile configuration. The stacking of the laminates are shown in Table 24.

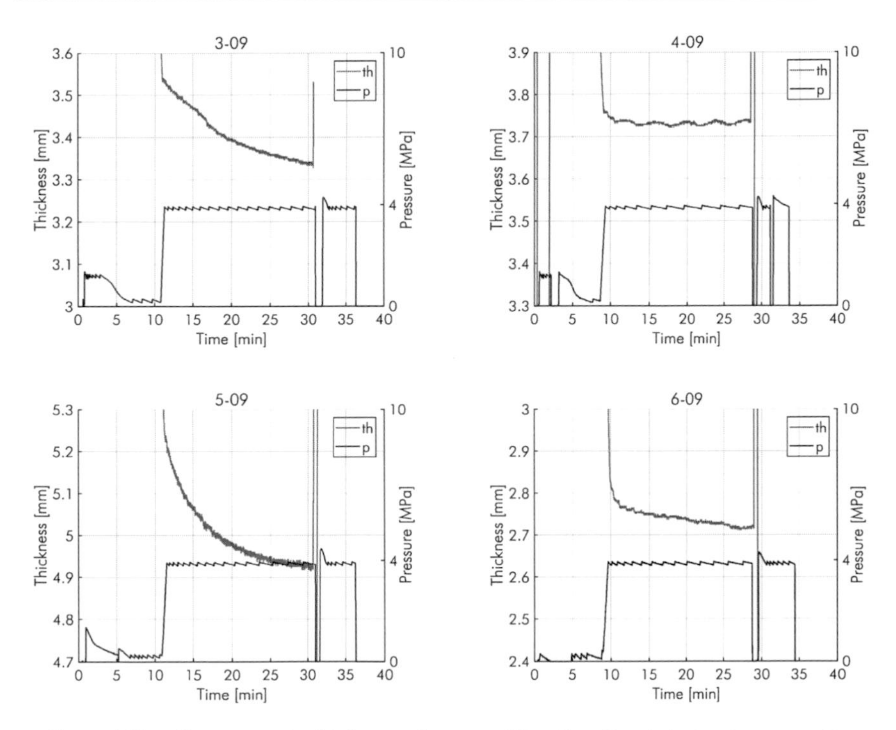

Figure 75 – Characteristic thickness change of Pre-3 to Pre-6 during impregnation

From the diagrams, it can clearly be seen that thickness decreases reciprocal to time, thus a qualitatively comparable characteristic as modelled in chapter 6.2.5. For Pre-3 and Pre-5, the decrease is more intensive than for other materials. Both textiles contain the same raw materials with PEEK 151G and a HTS45 P12 12k fibre. At the end of impregnation, thickness still decreases, which allows the assumption that impregnation would continue with more time resulting in smaller void contents. The change of thickness for Pre-4 and Pre-6 is smaller, which presumably is caused by the higher degree of mixing of the hybrid yarns inside the textile.

Figure 76 shows the relative thickness during the impregnation phase. For reasons of clarity, the bars showing the standard deviation are not shown. The data illustrate a decrease of thickness by 1% for Pre-4, 4% for Pre-6 and 6% for Pre-3 and Pre-5. Presumably, the smaller impregnation distances in Pre-4 and Pre-6 (see section 4.2) lead to a shorter required time for impregnation and therefore a lower compaction. However, since the specification of PEEK material is not known here, also a deviating viscosity could cause the smaller decrease of thickness.

Comparing the measured relative thickness in Figure 76 to the subsequently measured void content in Figure 77 shows correlating results. Focussing the measured void contents after 10 and 20 min in Figure 76, a reduction of the void content in dimensions of 1–2% is observed. These changes are comparable to the changes of the recorded thickness during processing. Here, the void content also decreases about 1–2% between 10 and 20 min of impregnation. Regarding the textile configurations, the thickness decreases reproducibly by approx. 4–5% after the first compression of the preform with its polymer in molten state. An exception is Pre-4 with only 1% compression after 10 min.

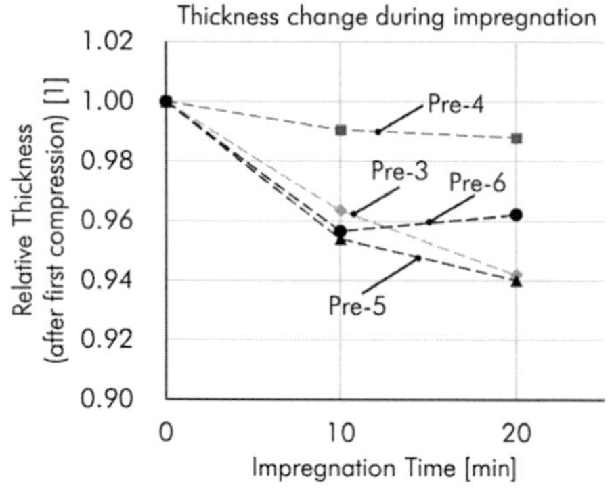

Figure 76 – Measured relative change of thickness during impregnation at 4 MPa

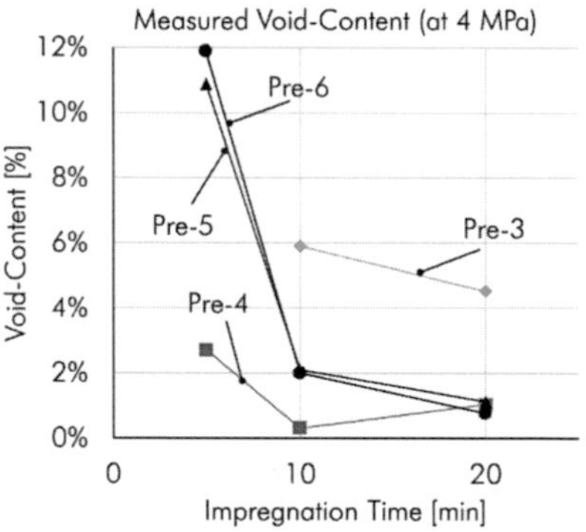

Figure 77 – Measured void content of laminates (pressure 4 MPa)

Conclusion

The results support the hypothesis that the thickness change during impregnation can give information about the current progress of impregnation. Thickness measurement can help to evaluate if the process of impregnation is ongoing or at its equilibrium. Furthermore, the results validate the assumption of an impregnation velocity which is reciprocal to time.

Nevertheless, since the results only cover relative thicknesses, the measurements cannot deliver information about the total degree of impregnation respective the void content. For this, absolute acquisition systems, probably containing one sensor each at both sides of the tooling, need to be considered.

8.2 Fibre-optic sensors

The objective of this chapter is to evaluate the time until the thermoplastic fibres start melting inside the preform. Knowledge about the time of these special occurrences helps for validating the models for heating. The acquisition is performed by means of fibre-optic sensors (FOS), which are embedded inside the preform.

Beside the determination of the phase of the polymer, the effect of the pressure drop during the Isoforming process and the corresponding transfer step is necessary to know. Therefore, the integrated fibre-optic sensors shall detect, if the pressure drop leads to delamination between fibre and polymer fluid.

Literature

Fibre-optic sensors allow analysing reflection spectra. A light-pulse, emitted by a light source outside the tooling setup, is sent through the glass fibre sensor and reflected at the open end of the fibre, which is inside the lay-up during processing. Depending on the ambient media, the so-called Fresnel reflection appears partially, depending on the difference of the index of refraction. The signal intensity, which is captured by a photo sensor, thus allows conclusions about changes of the media, in this case phase changes of the polymers fluid or solid phase. The capability to detect phase changes in consolidation processes was already validated in previous studies, dealing with compression moulding of CF-PEEK tapes or infusion processes [GAI15, ANT13]. An exemplary setup of integrated FOS is shown in Figure 78.

Measuring and experimental data

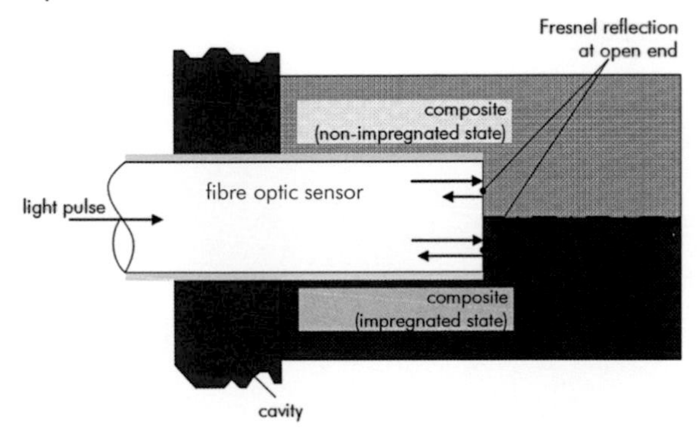

Figure 78 – Integrated fibre-optic sensor in laminate setup

The diameter of the fibre-sensor is 125 μm, leading to no significant disturbance of the impregnation process. To ensure an appropriate reflection, the end of the glass fibre needs to be cut off perpendicularly with high precision, which is done by a special guillotine. The reflected signal is then captured by a commercial software, Micron Electronics Enlight, and subsequently, the different wavelengths are interpreted by a MATLAB script.

For verification of the developed models in chapter 6 concerning the temperature distribution during heating and to verify the assumptions regarding the time for impregnation, the fibre-optic sensors were deposited at the surface of the stacking and in-between inner plies. In the experiment, four sensors were embedded into a Pre-5 preform, which contained 16 single plies. The positions of the sensors were in-between the cavity and ply 1(channel 1), ply 2 and ply 3 (channel 2), ply 4 and ply 5 (channel 3) and between ply 6 and ply 7 (channel 4), which is shown in Figure 79.

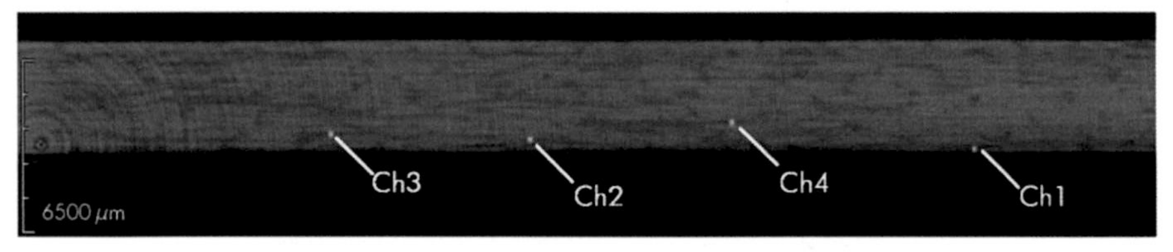

Figure 79 – CT-scan of laminate incl. embedded FOS

To ensure a complete impregnation, a pressure of 4 MPa and an impregnation duration of 20 min were chosen for the experiment.

Results and interpretation

The resulting temperature and pressure curves are displayed in Figure 80. Above the process diagram, the acquired signal strength in a wavelength range between 1510 and 1590 nm is illustrated. Here, a change of reflection intensity can be observed for the different stages of the consolidation process. From the point of pressure application, the signal strength immediately changes to approximately 50 dB. During the following heating phase, the reflection spectra remain without significant change. After 6–7 min, the measured temperature at the cavity (not inside the preform) increases above 350°C. The heating model, which was introduced in section 0, predicted a comparable heating trajectory. Regarding the temperature at the position of channel 4, the model and the experimentally investigated time for phase transition agree with a precision of 40 sec at this point. From around this temperature, a clear step of the reflection intensity can be observed, allowing the conclusion of a molten polymer at the sensor end.

During impregnation, no significant change of the intensity could be determined for channels 3 and 4. Channel 2, however, experiences a seriously fluctuating signal. Reason for this could be a damage of the sensor or an unsteady material at the fibres end, probably from moving voids. During the transfer step, an instant signal change could be observed, which may be a result from the current pressure drop. After transfer of the cavity and with beginning cooling phase, channel 3 shows no change of the reflection intensity. The signals in channel 4 and especially channel 2 clearly identify the cooling phase, which can be interpreted as a result of solidification or probably crystallisation. To prove this hypothesis, further experiments are necessary. Channel 1 showed unreasonable results, presumably resulting from failure during handling, which is the reason it is not shown here.

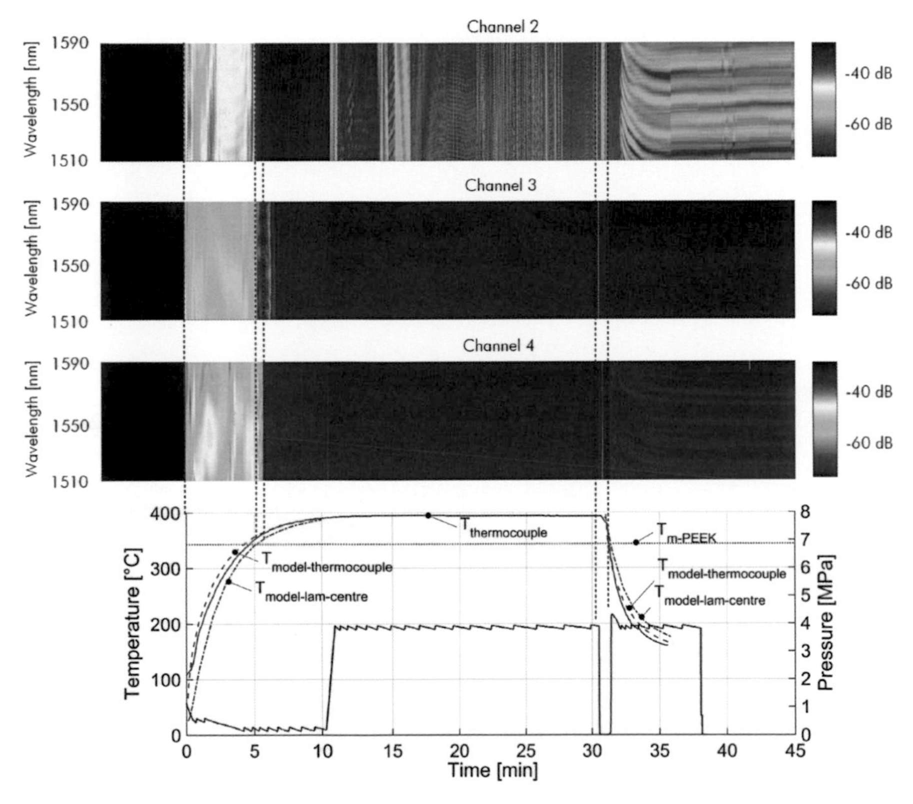

Figure 80 – Signals of three FOS during Isoforming-cycle

Conclusion

The experiments show that fibre-optic sensors allow a deeper look into the process of thermoforming. Changes in the materials phase are possible to detect, which can help to assure a state of homogeneous phase of the polymer inside the preforms throughout the whole thermoforming process in every point of the hybrid textile. This makes it specifically interesting for process development, model validation and temperature control. Concerning the heating characteristic, the phase of the thermoplastic material is almost equal across the laminate thickness during heating. If heating rates of the Isoforming process do not increase significantly, an equal phase state can be presupposed in the heating phase throughout the laminate thickness. The experiment validated the modelled heating trajectory from section 0. For evaluating the point of solidification, the experiments already show promising results. To investigate reliable data about the solidification point, certainly a closer look at the local temperatures and time steps with high temporal resolution would be necessary.

Considering the objective of this chapter, FOS helped to prove that the transfer step during Isoforming does not change the state of impregnation. A minor change could be determined in the reflection spectra during the pressure drop in the transfer step; nevertheless, a deconsolidation would have caused distinct changes in reflection intensity.

8.3 Transparent tooling

An obstacle for an in-situ process analysis for compression moulding at high temperatures are the metallic tooling components, which prevent a direct view into the cavity. A way to provide an insight into the process of consolidation is a compression tooling with single-side glass mould, which is developed here. The objective is to analyse impregnation and flow behaviour and possible other aspects to determine eventual improvements for the consolidation model and the manufacturing process.

Measuring and experimental data

The analysed specimens were analogue to Pre-3, consequently containing carbon and thermoplastic rovings in a side-by-side arrangement by TFP. However, due to limitations of the thermal conductivity of the glass tooling, thermoplastic fibres from polyamide 6 were processed. This enabled to reduce the required process temperature to 280°C, while the maximum applied pressure was 2,5 MPa. The processed material specification (B2703 by BASF) has a viscosity of ~ 270 Pa s and is thus comparable to the used PEEK polymers. The dimension of the specimens was 100 x 100 mm. The construction of the tooling contains a pressure plate made from borosilicate glass with a thickness of 60 mm. The specification SUPREMAX 33 by Schott provides a flexural strength of 25 MPa. In this construction, the glass block enables a resulting safety factor of 2,5 at 2,5 MPa compression pressure. With a temperature resistance up to 500°C, it would be applicable for consolidation experiments with PEEK theoretically. During impregnation, the process temperature was set to 280°C at 2 MPa applied pressure for 20 min.

Results

The experimental results show that as supposed macro-impregnation takes place within seconds after starting the impregnation step, while for micro-impregnation no effects could be seen visually. However, during impregnation, matrix flow in the direction of the shear-edge was detected. This flow mainly occured in lateral direction of the rovings. This in-plane flow clarified that the applied external pressure still acted on both the matrix system and the textile system.

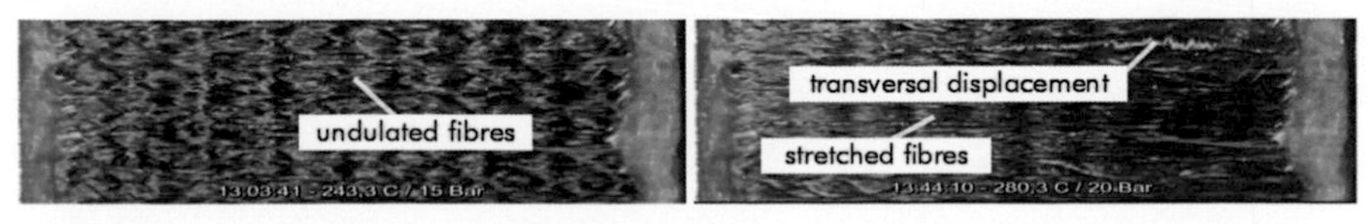

Figure 81 – Rovings after macro-impregnation and after stretching due to matrix flow

Since permeability is much higher in lateral direction of the roving than transverse (see equations (4.8) and (4.9)), the fluid tended to flow along the roving orientation. This caused shear stresses between the fluid and the fibre due to friction, which again stretched fibres to some extent (Figure 81). As conclusion, to a certain extent, matrix flow throughout the shear-edge has a positive effect on fibre stretching.

Since only one textile configuration was tested, no conclusions could be made with regard to commingled yarns. Anyway, it is a reliable assumption that while increasing the distribution homogeneity of thermoplastic and carbon fibres, as it is for commingled yarns, free spaces and resulting flow channels are scaled down. Consequently, the lateral flow might be reduced. This would again decrease the extent of stretched fibres by lateral flow.

In some experiments, an additional transversal matrix flow could be detected by a significant transversal realignment of some rovings (Figure 82). An explanation for the transversal flow is a slot thickness at the shear-edge, which is too large or distributed inhomogeneously. This shows the necessity to reduce the shear-edge slot to a minimum. Since the transversal flow did not occur in the beginning of the impregnation process, the assumption that it was not promoted by an inhomogeneous material distribution and resulting in-plane pressure gradients is supported.

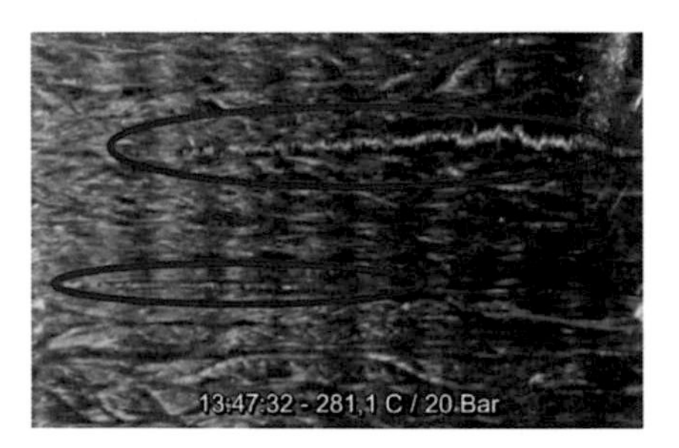

Figure 82 – Created gaps due to transverse flow during impregnation

Conclusion and error analysis

Concluding, the experiments show that the matrix flow in in-plane direction shall be considered if a comprehensive impregnation model should be developed. The differentiation between macro-impregnation and meso-impregnation was shown to be recommendable. Furthermore, the design of the tooling, especially the design of the shear-edge, takes a great influence on the impregnation quality, as well as the homogeneity of the arrangement of the fibre materials. A further way to prevent fibre realignment could be sewing yarns with higher thermal stability. For PEEK matrices, these yarns could consist of aramid (not para-aramid) glass or metal, for instance.

9 Conclusion and outlook

This thesis had the intention to give a comprehensive overview about the influence of material and process parameters on the structure of the resulting thermoplastic composite parts made from hybrid textiles. Namely the polymer raw materials, so as different textile configurations were investigated. A process model furthermore deviated recommendations for process conditions for consolidation. A further objective was to develop a cost-efficient production process for thermoforming of net-shape components from high-performance thermoplastic composites.

Conclusion – polymers

Concerning the analysis of thermoplastic materials, two specifications of PEEK were investigated: a classic specification with VICTREX PEEK 151G and a new launched low-melt PEEK AE-250™-polymer, which is promising especially for overmoulding applications. Both polymers were analysed concerning their thermal behaviour during thermoforming, which allows a detailed modelling of the heating and cooling trajectories and the impregnation behaviour with the corresponding effects. Crystallisation and the temperature-dependent mechanical properties CTE and flexural storage modulus were examined, so that adequate input parameters were provided for a detailed consolidation model.

Conclusion – hybrid textiles

Together with carbon fibre, the thermoplastic fibres are combined in hybrid textiles. Four different textile configurations were analysed regarding their thermal and compression characteristics. Furthermore, the textiles were analysed in terms of their flow path length, resulting from the thickness of the carbon fibre plies. The obtained data illustrate the diversity of the different textiles, resulting from different textile raw materials and the textile manufacturing processes (TFP and NCF-manufacturing). Beside flow path length, also the homogeneity of the distribution distinguished the textiles. While a ply-wise arrangement of thermoplastic and carbon fibres leads to homogeneous but long flow path lengths, pre-mixing the fibres to hybrid yarns helps in reducing the required penetration distance.

The investigated values for the thermal conductivity in the unimpregnated and the impregnated state were figured out experimentally or by modelling. They allow a prediction of the temperature distribution during compression moulding.

A forecast of the impregnation characteristic requires information about the compression behaviour of the textile. This was investigated by a newly developed porous tooling which minimises the influence of the integrated polymer in its molten state. The results showed different stress–FVC characteristics for different textiles, which were used to develop predictive models. However, the results also demonstrated that for a first compression of the textile, the influence of the molten polymer cannot be neglected. Only after a first compression, a reproducible behaviour for compression and relaxation was observed.

Conclusion – Isoforming process

Beside consolidation analysis, an objective of this thesis was to develop a fast process for the consolidation of hybrid textiles with high-temperature thermoplastics. With the Isoforming process, a suitable method was realised for thermoforming of 2,5D net-shape components from hybrid textiles with increased heating and cooling rates. Compared to classic variothermal moulds for temperatures up to 400°C, no hazardous oil for temperature control is required with this method. Beside time, the required energy is reduced drastically, since most of the components remain at an isothermal temperature. Additionally, the process provides a high variability concerning differently shaped components. A new shape only requires a new cavity system without active temperature management system and minor adjustment of the compression toolings. A further improvement concerning impregnation quality could result from evacuating the cavity.

Conclusion and outlook – model development

In order to understand the process of consolidation, a model for heating, impregnation and cooling was developed. For heating, a good agreement of the modelled time until crossing the melting temperature was demonstrated compared to experiments. For cooling, concerning the transient tooling temperatures, the modelled values also agreed satisfactory with the measured data. The impregnation model focussed on the penetration of the polymer fluid into the carbon fibres on meso-scale. Here the model provided appropriate information about the sensitivity of material and process parameters on the impregnation process. Especially, the modelled void content was in good agreement with the experimentally measured values. Here the applied pressure was identified to have a dominating influence on the prevention of voids. The results generally recommend high impregnation pressures in ranges of 4 MPa or above to minimise voids. However, for the impregnation time, the results only allow a qualitative interpretation. In this context, the flow path length has a governing influence. For improvement, a multi-scale impregnation model, which also considers the macro-scale beside the meso-scale including a focus on all simplifications is recommended. Also the influence of the sizing of the carbon fibres on friction shall be considered in future analyses.

Conclusion – sensitivity analysis

In order to validate the impregnation model and to investigate impregnation and its influence on mechanical properties in detail, consolidated laminates were manufactured by the Isoforming process. Impregnation time and applied pressure were modified for all four investigated textiles. As in the impregnation model before, the pressure was evaluated to be the governing parameter for the reduction of voids. Concerning the required time for impregnation, the flow path length in different textiles had a major influence. Generally, durations for impregnation from and above 10 min delivered eligible results. A subsequent analysis demonstrated the dependence of the inter-laminar shear strength of the void content. Here porosities below 2% were identified to be recommendable.

Conclusion – process analysis

Finally, an analysis of the consolidation process investigated methods for quality assurance and to obtain a closer look into the phenomena inside the cavity during consolidation. The results of the online thickness measurement indicated that it can help to evaluate the progress of impregnation. Absolute thickness values could be obtained with more than one sensor integrated into the tooling. An in-situ integration of fibre-optic sensors allowed a deeper look inside the process of thermoforming and the validation of the thermal process model. Changes in the materials phase were feasible to detect, which helped to validate a homogeneous temperature distribution inside preforms throughout the whole thermoforming process. To gain more knowledge about the polymer flow inside the cavity during impregnation, a glass tooling provided insight. The experiments show that the matrix flow in in-plane direction and corresponding fibre–fluid interaction should be taken into account for a comprehensive impregnation model. The results support the differentiation between macro- and meso-impregnation.

Outlook – further improvement of hybrid textiles and their concurrence situation

Concluding, this thesis contributed to the knowledge about the internal phenomena during the consolidation process of hybrid textiles. Feasible methods for the consolidation of high-performance thermoplastic composite components were developed including the elaboration of their advantages and limitations. In order to provide high-quality components, it was shown that homogeneous pre-mixing of the fibres is necessary. Of course, the stress field between mixing, damaging of the carbon fibres and resulting material costs needs to be considered. The material is in concurrence to tape-laying processes and will only be applied, if a high quality can be achieved combined with the exploitation of their in-plane and out-of-plane drapeability and the feasibility of their alignment according to the dominant load path directions.

If the demand for complex shaped components, which require the drapeability of a textile on the one hand and the direct transfer of loads on the other hand, increases, hybrid textiles are a promising material class. This thesis contributed to a better understanding of the materials and demonstrated ways to exploit their potentials.

Literature

ADV10 Advani, S., Sozer, E. et al: Process Modelling in Composites Manufacturing, Second Edition, CRC Press, Abingdon, 2010

AGK18 AGK GmbH: http://www.agk.de/html/isolierwerkstoffe/k_therm_as.htm, access on 22.01.2018

ANT13 V. Antonucci, M. Giordano, L. Nicolais, A. Calabrò, A. Cusano, A. Cutolo, S. Inserra: Resin flow monitoring in resin film infusion process, Journal of Materials Processing Technology, Vol. 143–144, 2003

ARE04 Areerat, S., Funami, E., Hayata, Y., Nakagawa, D., Ohshima, M.: Measurement and Prediction of Diffusion Coefficients of Supercritical CO2 in Molten Polymers, Polymer Engineering and Science, Vol. 44, 2004

ARH16 Arrhenius, S.: The viscosity of pure liquids, Meddelanden Från K. Vetenskapsakademiens, Nobelinstitut, Vol. 3, Stockholm, 1916

BER99 Bernet, N., Michaud, V., Bourban, P.-E., Månson, J.-A. E: An impregnation model for consolidation of thermoplastic composites made from commingled yarns, Journal of Composite Materials, Vol. 33, 1999

BOS14 Bostan, L., Schiebel, P.: Funktionalisierte Fasern zur Thermofixierung von PEEK / CF-Preforms für Hochleistungsfaserverbundbauteile: BiKo-PEEK Tow Placement, Book on Demand Verlag, Norderstedt, 2014

BOW16 Bouwman, M., Donderwinkel, T., Wijskamp, S.: Overmoulding - An Integrated Design Approach for Dimensional Accuracy and Strength of Structural Parts, ITHEC 2016, Bremen, 2016

BOU01 Bourban, P.-E., Bernet, N., Zanetto, J.-E., Månson: Material phenomena controlling rapid processing of thermoplastic composites, Composites: Part A, Vol. 32, 2001

BRA13 Brauner, C.: Analysis of process-induced distortions and residual stresses of composite structures, PhD-Thesis, Faserinstitut Bremen, Logos Verlag Berlin, 2013

CAR37 Carman, P.: Fluid flow through granular beds, Institution of Chemical Engineers, London, 1937

CHI94 Chan, T.W., Isayev, A.I.: Quiescent polymer crystallization: Modelling and Measurements, Polymer Engineering and Science, Vol. 34, 1994.

CHO90 Choy, C.L., Leung, W.P.: Thermal expansion of Poly(ether-ether-ketone) (Peek), Journal of Polymer Science: Part B: Polymer Physics, Vol. 28, 1990

CHO94 Choy, C.L., Kwok, K.W., Leung, W.P., Lau, F.P.: Thermal conductivity of poly(ehter ether ketone) and its short-fiber composites, Journal of Polymer Science: Part B: Polymer Physics, Vol. 32, 1994

CON95 Connor, M, Toll, S., Manson, J.: On surface energy effects in composite impregnation and consolidation, Composites Manufacturing, Vol. 6, 1995

DIE77 Dietz, M.: Die Wärme- und Temperaturleitfähigkeit von Kunststoffen, Colloid & Polymer Science 255, Dr. Dietrich Steinkopff Verlag, Darmstadt, 1977

EBN13 Ebnesajjad, S., Ebnesajjad, E.: Surface Treatment of Materials for Adhesive Bonding, Elsevier Science & Technology, 2013

EHL01 Ehleben, M.: Herstellung von endlosfaserverstärkten Rohren mit thermoplastischer Matrix im Schleuderverfahren, PhD-Thesis, Technische Universität Darmstadt, Darmstadt, 2001

EHR03 Ehrenstein, G.: Praxis der Thermischen Analyse von Kunststoffen, Hanser Fachbuch, Munich, 2003

EMB07 Emberey, C., Milton, N., Berends, J., van Tooren, M.: Application of Knowledge Engineering Methodologies to Support Engineering Design Application Development in Aerospace, 7th AIAA Aviation Technology, Integration and Operations Conference (ATIO), Belfast, 2007

FIC55 Fick, A.: Ueber Diffusion, Annalen der Physik, vol. 170, Issue 1, WILEY-VCH Verlag GmbH & Co. KGaA, Weinheim, 1855

FOU95 Fourné, F.: Synthetische Fasern - Herstellung, Maschinen und Apparate, Eigenschaften, Hanser Fachbuch, Munich, 1995

FRI97 Friedrich, K., Hou, M., Krebs, J: Thermoforming of continuous fibre/thermoplastic composite sheets, Composite Materials Series, Vol. 11, 1997

GAI15 Gaitzsch, R., Koerdt, M., Brauner, C., Kroll, L., Herrmann, A.: Signal evaluation of fibre optical sensors embedded between unidirectional thermoplastic prepreg tapes in a hot-press consolidation for online process monitoring, ICCM20, Copenhagen, 2015

GEB91 Gebart, B.R.: Permeability of Unidirectional Reinforcements for RTM, Journal of Composite Materials, Vol. 26, 1991

GLO13 Glowania, M.: Untersuchung und Methodenentwicklung zur Steigerung der Wärmeleitfähigkeit von Faserverbundkunststoffen, PhD-Thesis, ITA Aachen, Shaker Verlag, Herzogenrath, 2013

GUT87 Gutowski, T.G. et al: Consolidation Experiments for Laminate Composites, Journal of Composite Materials, Vol. 21, 1987

HA97 Ha, S.-W., Hauert, R., Ernst, K.-H., Wintermantel, E.: Surface analysis of chemically-etched and plasma-treated polyetheretherketone (PEEK) for biomedical applications, Surface and Coatings Technology, Vol. 96, 1997

HAF98 Haffner, S.M., Friedrich, K., Hogg, P.J., Busfield, J.: Finite Element Assisted Modelling of the Microscopic Impregnation Process in Thermoplastic Preforms, Applied Composite Materials, Vol. 5, 1998

HEN03 Henry, W.: Experiments on the quantity of gases absorbed by water, at different temperatures, and under different pressures, Philosophical Transactions of the Royal Society of London, Vol. 93, 1803

HOP14 Hopmann, C. et al: Neue Prozesskette für Faserverstärkte Thermoplaste, wt Werkstattstechnik online, Vol. 9-2014, Springer-VDI-Verlag, Düsseldorf, 2014

INC96 Incropera, F., DeWitt, D.: Fundamentals of Heat and Mass Transfer, John-Wiley & Sons, New York, 1996

JAR12 Jaroschek, C.: Das Ende des Biegemoduls, Journal of Plastics Technology, Vol. 8, Carl Hanser Verlag, Munich, 2012

KAS10 Kastner, J., Plank, B., Salaberger, D., Sekelja, J.: Defect and Porosity Determination of Fibre Reinforced Polymers by X-ray Computed Tomography, 2nd International Symposium on NDT in Aerospace 2010, Hamburg, 2010

KIM91 Kim, Y.R., McCarthy, S.P., Fanucci, J.P.: Compressibility and relaxation of fiber reinforcements during composite processing, Polymer Composites, Vol. 12, 1991

KM16 Krauss Maffei: T-RTM fit für die Großserie, Pressemitteilung zur K2016, https://www.kraussmaffei.com/imm-de/presse/d/k2016_t-rtm.html, access on 22.01.2018

KOE16 Koerdt, M., Schiebel, P., Focke, O., Herrmann, A.: Hybrid Textiles – The other way of forming high-performance thermoplastic composites for primary structure, ECCM17, Munich, 2016

KOE18 Koerdt, M., Focke, O., Herrmann, A.: Impregnation analysis of compression moulding of thermoplastic composites made from hybrid textiles by thickness assessment and optical analysis methods, Symposium Lightweight Design in Product Development, Zurich, 2018

KOZ27 Kozeny, J.: Über kapillare Leitung des Wassers im Boden, Sitzungsberichte der Kaiserlichen Akademie der Wissenschaften, Vienna, 1927

KUE16 Kühn, F. et al: Ceramic pressing tool for variothermal processing of thermoplastic fiber composites, ECCM17, Munich, 2016

LI15 Li, B., Deleglise-Lagardere, M., Park, C., Lacrampe, M.: Analysis of intra-yarn impregnation in commingled yarn thermoplastic composites consolidation process, ICCM20, Copenhagen, 2015

LON01 Long, A.C., Wilks, C.E., Rudd, C.D.: Experimental characterisation of a commingled glass/polypropylene composite, Composites Science and Technology, Vol. 61, 2001

LU04 Lu, M., Ye, L., Mai, Y.: Thermal de-consolidation of thermoplastic matrix composites -II. "Migration" of voids and "re-consolidation", Composites Science and Technology, Vol. 64, 2004

MAN89 Månson, J-A., Seferis, J.: Autoclave Processing of PEEK/Carbon Fiber Composites, Journal of Thermoplastic Composite Materials, Vol. 2, 1989

MAY18 May, D., Aktas, A., Yong, A.: International benchmark exercises on textile permeability and compressibility characterizsation, ECCM18, Athens, 2018

MER10 Merotte, J., Simacek, P., Advani, S.G.: Resin flow analysis with fiber preform deformation in through thickness direction during Compression Resin Transfer Molding, Composites: Part A, Vol. 41, 2010

MIC01 Michaud, V., Mortensen, A.: Infiltration processing of fibre reinforced composites - governing phenomena, Composites: Part A, Vol. 32, 2001

MIC97 Michaud, V., Sommer, J., Mortensen, A.: Infiltration of fibrous preforms by a pure metal: Part V. Influence of preform compressibility, Metallurgical and Materials Transactions A, Vol. 30, 1999

MUZ97 Muzzi, J., Colton, J.: The Processing Science of Thermoplastic Composites, Chapter from Advanced Composites Manufacturing, Gutowski T. et. al, John Wiley & Sons, 1997

NAK73 Y.P. Nakamura, K. Katayama, and T. Amano. Some aspects of non–isothermal crystallization of polymers. II. Consideration of the isokinetic condition, Journal of Applied Polymer Science, Vol. 17, 1973

NAS97 Ohlhorst, C. W., Vaughn, W. L., Ransone, P. O., Tsou, H. T.: NASA Technical Memorandum 4787 - Thermal Conductivity Database of Various Structural Carbon-Carbon Composite Materials, Langley Research Center, Hampton, 1997

NIL13 Nilsson, F., Hallstensson, K., Johansson, K., Umar, Z., Hedenqvist, M.: Predicting Solubility and Diffusivity of Gases in Polymers under High Pressure: N2 in Polycarbonate and Poly(ether-ether-ketone), Industrial & Engineering Chemistry Research, Vol. 52, 2013

OSW15 Osswald T., Rudolph N.: Polymer Rheology: Fundamentals and Applications, Carl Hanser Verlag, Munich, 2015

REG17 Regis, M., Bellare, A., Pascolini, T., Bracco, P.: Characterization of thermally annealed PEEK and CFR-PEEK composites: Structure-properties relationships, Polymer Degradation and Stability, Vol. 136, 2017

RIJ07 Van Rijswik, K.: Thermoplastic composites wind turbine blades: vacuum infusion technology for anionic polyamide-6 composites, PhD-Thesis, TU Delft, Delft, 2007

ROL95 Rolfes, R., Hammerschmidt, U.: Transverse thermal conductivity of CFRP laminates: A numerical and experimental validation of approximation formulae, Composites Science and Technology, Vol. 54, 1995

ROU13 Rouhi, M., Wysocki, M., Larsson, R.: Modeling of coupled dual-scale flow–deformation processes in composites manufacturing, Composites Part A: Applied Science and Manufacturing, Vol. 46, 2013

ROU15 Rouhi, M., Wysocki, M., Larsson, R.: Experimental assessment of dual-scale resin flow-deformation in composites processing, Composites Part A: Applied Science and Manufacturing, 2015

SAU18 https://www.stauberstahl.com/werkstoffe/12343-werkstoff-datenblatt/, access on 22.01.2018

SCA17 Schaal, L.: A high performance and cost-effective solution for automotive and aeronautical applications, Reinforced Plastics, 2017

SCH07 Schürmann, H.: Konstruieren mit Faser-Kunststoff-Verbunden, Springer, Heidelberg, 2007

SCH16 Schäfer, P.M.: Experimental Investigation of inter-layer thermal contact resistance and its relevance for consolidation of thermoplastic composites, ECCM17, Munich, 2016

SCH18 Schiebel, P.: Entwicklung von Hybrid-Preforms für belastungsgerechte CFK-Strukturen mit thermoplastischer Matrix, PhD-Thesis, Faserinstitut Bremen, Logos Verlag Berlin, 2018

SCO12 Scholl, S.: Zur kontinuierlichen Herstellung prismatischer Leichtbauprofile aus Faser-Kunststoff-Verbunden mit thermoplastischer Matrix, PhD-Thesis, Technische Universität Darmstadt, Shaker Verlag, 2012

SCU11 Schuck, M.: Kunststoffe als Leichtbauwerkstoffe, wissenschaftstag metropolregion nürnberg, Nürnberg, 2011

SCW14 Schwing, B., Kaschel, S., Brok, W.: New Concepts for Structure Parts Based on Short Fibre Reinforced Injection Molding, ITHEC 2014, Bremen, 2014

SHI16 Shi, H., Fernandez Villegas, I., Bersee, H: Analysis of void formation in thermoplastic composites during resistance welding, Journal of Thermoplastic Composite Materials, Vol. 30, 2016

SKO16 Skowronek, Marco: Entwicklung eines gläsernen Presswerkzeugs zur optischen Untersuchung des Imprägnierverhaltens hybrider Textilien, Master Thesis, Faserinstitut Bremen, Bremen, 2016

SOM92 Sommer, S.L.: Infiltration of deformable porous media, doctoral thesis, Massachusetts Institute of Technology, Department of Materials Science and Engineering, Boston, 1992

STE12 Sterk, S.: Pressed Formed Thermoplastic Window Frames, International Symposium for Composites Manufacturing ISCM 2012, Braunschweig, 2012

STO16 Stokes-Griffin, C.M., Compston, P.: Investigation of sub-melt temperature bonding of carbon-fibre/PEEK in an automated laser tape placement process, Composites: Part A, Vol. 84, 2016

STU16 Studer, J., Dransfeld, C., Fiedler, B.: Direct thermoplastic melt impregnation of carbon-fibre fabrics by injection moulding, ECCM17, Munich, 2016

TAL87 Talbott, M.F., Springer, G.S., Berglund, L.A.: The Effects of Crystallinity on the Mechanical Properties of PEEK Polymer and Graphite Fiber Reinforced PEEK, Journal of Composite Materials, Vol. 21, 1987

THO03 Thomann, I.: Direct stamp forming of non-consolidated carbon\thermoplastic fibre commingled yarns, PhD-Thesis, ETH-Zürich, Zurich, 2003

TOL95 Toll, S., Manson, J.-A.: Elastic Compression of a Fiber Network, Journal of Applied Mechanics, Vol. 62, 1995

TOL98 Toll, S.: Packing mechanics of fiber reinforcement, Polymer Engineering and Science, Vol. 38, 1998

VIC12 Victrex PEEK 151G datasheet, Victrex.com, access on 10.04.2012

VOY63 Voyutskii, S.S.: Autohesion and adhesion of high polymer, Polymer Review, Vol. 4, New York, 1963

WAK98 Wakeman, M.D., Cain, T.A., Rudd, C.D., Brooks, R., Long, A.C.: Compression moulding of glass and polypropylene composites for optimised macro- and micro-mechanical properties – 1 commingled glass and polypropylene, Composite Science and Technology, Vol. 58, 1998

WAN18 Wang, Y., Chen, B., Evans, K., Ghita, O.: Enhanced Ductility of PEEK thin film with self-assembled fibre-like crystals, Scientific Reports, Vol. 8, Article number: 1314, 2018

WIJ05 Wijskamp, S.: Shape distortions in composites forming, PhD-Thesis, University of Twente, Enschede, 2005

WYS05 Wysocki, M., Larsson, R., Toll, S.: Hydrostatic consolidation of commingled fibre composites, Composite Science and Technology, Vol. 65, 2005

WYS07 Wysocki, M., Toll, S., Larsson, R.: Press forming of commingled yarn based composites - the preform contribution, Composites Science and Technology, Vol. 67, 2007

YAN02 Yang, F., Pitchumani, R.: Healing of Thermoplastic Polymers at an Interface under Nonisothermal Conditions, Macromolecules, Vol. 35, 2002

YAN13 Yang, Y., Robitaille, F., Hind, S.: Thermal conductivity of carbon fiber fabrics, ICCM19, Montreal, 2013

Appendix – Numerical simulation of impregnation

The model and the further numerical approach for solution is mainly based on previous works of Sommer, Michaud, Jespersen and Wysocki. [SOM92, MIC97, MIC01, WYS05, JES08]

Transformation

To simplify the system of partial differential equations, a similarity variable is introduced, which combines the two independent variables of time and position in one single variable χ by a Boltzmann transformation. This transforms the problem in terms of ordinary differential equations, so that:

$$u(x,t) \rightarrow u(\chi)\text{ , with } \chi = \frac{z}{t^{1/\beta}} \text{ as similarity variable with } \beta = 2\,.$$

In this context, χ expresses a non-dimensional position in the impregnated region and is defined by:

$$\chi = \frac{z - z_e}{\Psi\sqrt{t}} = \frac{z - z_e}{z_f - z_e}.$$

Here z_e describes the interface between neat matrix and impregnated preform in relaxed state and z_f as the position of the infiltration front. All governing equations are transformed and need to be expressed by χ.

Furthermore, Ψ is a scalar value which defines a flow rate, which is chosen initially so that $\chi = 1$ at the flow front. The impregnated length is thus expressed by:

$$\chi(L) = 1 = \frac{L}{(\Psi\sqrt{t})} \quad \rightarrow \quad L = \Psi\sqrt{t}\,.$$

Deriving χ with respect to time and position yields:

$$\frac{\partial\chi}{\partial t} = -\frac{z - z_e}{2\Psi t^{3/2}} - \frac{u_F(x = 0)}{\Psi\sqrt{t}} = -\frac{\chi}{2t} - \frac{u_F(\chi = 0)}{\sqrt{t}}$$

$$\frac{\partial\chi}{\partial z} = \frac{1}{\Psi\sqrt{t}}.$$

Assumptions

Necessary requirement for the validity of transformation is that no changes of material parameters (e.g. viscosity) or a change of pressure occurs inside a control volume during a discretised time step. Reactions inside the control volume need to be fast, so that an equilibrium of forces is fulfilled anytime.

Rewriting the governing equations

Substitution of dz in Darcy´s equation (6.7) with $d(z - z_e) = d(\chi\,\Psi\,\sqrt{t})$ yields:

$$u_M - u_F = \frac{K}{\eta(1-\varphi_F)}\,\frac{d\varphi_F}{d\chi}\,\frac{1}{\Psi\,\sqrt{t}}\,\frac{d\sigma}{d\varphi_F}.$$

Rewriting the velocity equations (impregnation front and preform front)

Further, the velocities of the penetrating polymer fluid and the relaxing preform are expressed in terms of χ, and considering the introduced equations for the liquid and solid positions $l(\chi)$ and $s(\chi)$:

$$u_M(\chi,t) = \frac{\Psi l(\chi)}{2\sqrt{t}} \tag{A.1}$$

$$u_F(\chi,t) = \frac{\Psi s(\chi)}{2\sqrt{t}}. \tag{A.2}$$

Now rewriting the equations for continuity (eqs. (6.2) and (6.3)) delivers:

$$\frac{\partial \varphi_F}{\partial t} + \frac{\partial}{\partial z}(\varphi_F u_F) = 0$$

with

$$\frac{\partial \varphi_F}{\partial t} = \frac{\partial \varphi_F}{\partial \chi}\frac{\partial \chi}{\partial z} = \frac{\partial \varphi_F}{\partial \chi}\left(-\frac{\chi}{2t} - \frac{u_s(\chi = 0)}{\Psi\sqrt{t}}\right)$$

$$\frac{\partial \varphi_F}{\partial z} = \frac{\partial \varphi_F}{\partial \chi}\frac{\partial \chi}{\partial z} = \frac{\partial \varphi_F}{\partial \chi}\left(\frac{1}{\Psi\sqrt{t}}\right)$$

$$\frac{\partial u_F}{\partial z} = \frac{\partial u_s}{\partial \chi}\frac{\partial \chi}{\partial z} = \frac{\partial(\Psi s(\chi)/2\sqrt{t})}{\partial \chi}\frac{\partial \chi}{\partial z} = \frac{1}{2t}\frac{\partial s(\chi)}{\partial \chi}$$

leads to:

$$\frac{ds}{d\chi} = \left(-\chi - \frac{2u_s(x = 0)\sqrt{t}}{\Psi} + \frac{2u_s\sqrt{t}}{\Psi}\right)\left(-\frac{d\varphi_F/d\chi}{\varphi_F}\right)$$

and

$$-\frac{\partial \varphi_F}{\partial t} + \frac{\partial((1-\varphi_f)u_M)}{\partial z} = 0$$

with

$$\frac{\partial \varphi_F}{\partial t} = \frac{\partial \varphi_F}{\partial \chi}\frac{\partial \chi}{\partial z} = \frac{\partial \varphi_F}{\partial \chi}\left(-\frac{\chi}{2t} - \frac{u_F(\chi = 0)}{\Psi\sqrt{t}}\right)$$

$$\frac{\partial(1-\varphi_F)}{\partial z} = \frac{\partial(1-\varphi_F)}{\partial \chi}\frac{\partial \chi}{\partial z} = \frac{\partial(1-\varphi_F)}{\partial \chi}\left(\frac{1}{\Psi\sqrt{t}}\right)$$

$$\frac{\partial u_M}{\partial z} = \frac{\partial u_M}{\partial \chi}\frac{\partial \chi}{\partial z} = \frac{\partial\left(\Psi l(\chi)\,/\,2\sqrt{t}\right)}{\partial \chi}\frac{1}{\Psi\sqrt{t}} = \frac{1}{2t}\frac{\partial l(\chi)}{\partial \chi}$$

leads to:

$$\frac{dl}{d\chi} = \left(\chi + \frac{2u_F(\chi = 0)\sqrt{t}}{\Psi} + \frac{2u_M\sqrt{t}}{\Psi}\right)\left(-\frac{d\varphi_F/d\chi}{1-\varphi_F}\right).$$

Furthermore,

$$-\frac{\partial \varphi_F}{\partial t}+\frac{\partial((1-\varphi_F)u_M)}{\partial z}=\frac{\partial \varphi_F}{\partial t}+\frac{\partial}{\partial z}(\varphi_F u_F)=0$$

leads to: $$\frac{\partial((1-\varphi_F)l+\varphi_F s)}{\partial \chi}=0\,.$$

Summarising, the governing equations can be expressed with respect to χ.

Darcy

$$\frac{d\varphi_F}{d\chi}=\frac{(l(\chi)-s(\chi))(1-\varphi_F(\chi))\,\eta\,\Psi^2}{2K(\varphi_F)\,\frac{d\sigma}{d\varphi_F}} \tag{A.3}$$

Conservation of mass

Solid $$\frac{ds}{d\chi}=\big(s(\chi)-\chi-s(0)\big)\left(-\frac{d\varphi_F\,/d\chi}{\varphi_F(\chi)}\right) \tag{A.4}$$

Liquid $$\frac{dl}{d\chi}=\big(l(\chi)-\chi-s(0)\big)\left(\frac{d\varphi_F\,/d\chi}{1-\varphi_F(\chi)}\right). \tag{A.5}$$

The determination of l can be simplified due to a summation of equations (A.3) and (A.4). After integration between χ and 1, the sum yields:

$$l(\chi)=\frac{\big(1-\varphi_{Fc}\big)\big(1+s(0)\big)-\varphi(\chi)\,s(\chi)}{1-\varphi(\chi)}. \tag{A.6}$$

Consequently, impregnation can be described by two ordinary differential equations (A.3) and (A.4) and the dependent equation (A.6).

Boundary conditions

Two boundary conditions are required to solve the ODE system. At the interface between dry preform and infiltration front, conditions for the position of the textile and the fibre volume content are known. The fibre volume content at the interface is known from the stress–strain relation of the preform in the compressed state from Section 4.4:

$$\varphi(\chi = 1) = \varphi_{Fc}$$

If the capillary pressure drop is neglected, the boundary condition for $s(1)$, representing the preform position at the impregnation front simplifies to:

$$s(\chi = 1) = 0\,.$$

Furthermore, the local fibre volume content at the relaxed preform front is known from the stress–relaxation relation of the textile:

$$\varphi_F\,(\chi = 0) = \varphi_{Fr}\,\left(@\,\sigma = \left(p_{app} - p_{Gas}\right) \cdot 0.01\right).$$

In the model, φ_{Fr} is defined as the fibre volume content, where the locally applied stress on the textile is only 1% of the maximum acting matrix pressure at the infiltration front.

Initial conditions

An additional value for s at $\chi = 0$ needs to be known for solving the ODE system. Furthermore, the introduced flow rate Ψ is required. If process parameters are stationary, Ψ remains unchanged. Both values are guessed initially and evaluated by comparing the results with the boundary conditions at $\chi = 1$.

Re-transformation

For re-transformation, the determined data for Ψ, $s(\chi)$ and $l(\chi)$ are used for calculating the local and time-dependent velocities u_M and u_F with equations (A.1) and (A.2).

Subsequently, the impregnated distance can be determined by integration with respect to time. The local fibre volume content is a result of the ODE system. From these data, further information can be gathered, namely the pressure drop, the already penetrated polymer respective the impregnated fibre volume, the degree of impregnation and the overall void content:

$$\Delta p(t) = \sum_{\chi=0}^{\chi=1} \frac{\big(l(\chi) - s(\chi)\big) \cdot (1 - \varphi_F\,(t,\chi)) \cdot \eta\Psi^2}{-2 \cdot K(\chi)} \cdot \Delta\,\chi \tag{A.7}$$

$$Vol_{M-impr}(t) = \sum_{z=z_e}^{z=z_f} (1 - \varphi_F\,(t,z)) \cdot dz(z) \tag{A.8}$$

$$Vol_{CF-impr}(t) = \sum_{z=z_e}^{z=z_f} \varphi_F\,(t,z) \cdot dz(z) \tag{A.9}$$

$$DoI(t) = \frac{Vol_{CF-impr}(t)}{th_{CF}(t=0) \cdot \varphi_{Fc}} \tag{A.10}$$

$$vc(t) = \frac{th_{CF}(t) \cdot \big(1 - \varphi_{Fc}\big)}{th_{CF}(t) + Vol_{CF-impr}(t) + Vol_{M-impr}(t)}. \tag{A.11}$$

Iteration

For iteration, a tailored scheme was developed for adjusting both values Ψ and s independently from each other. The sequence of the model is illustrated below:

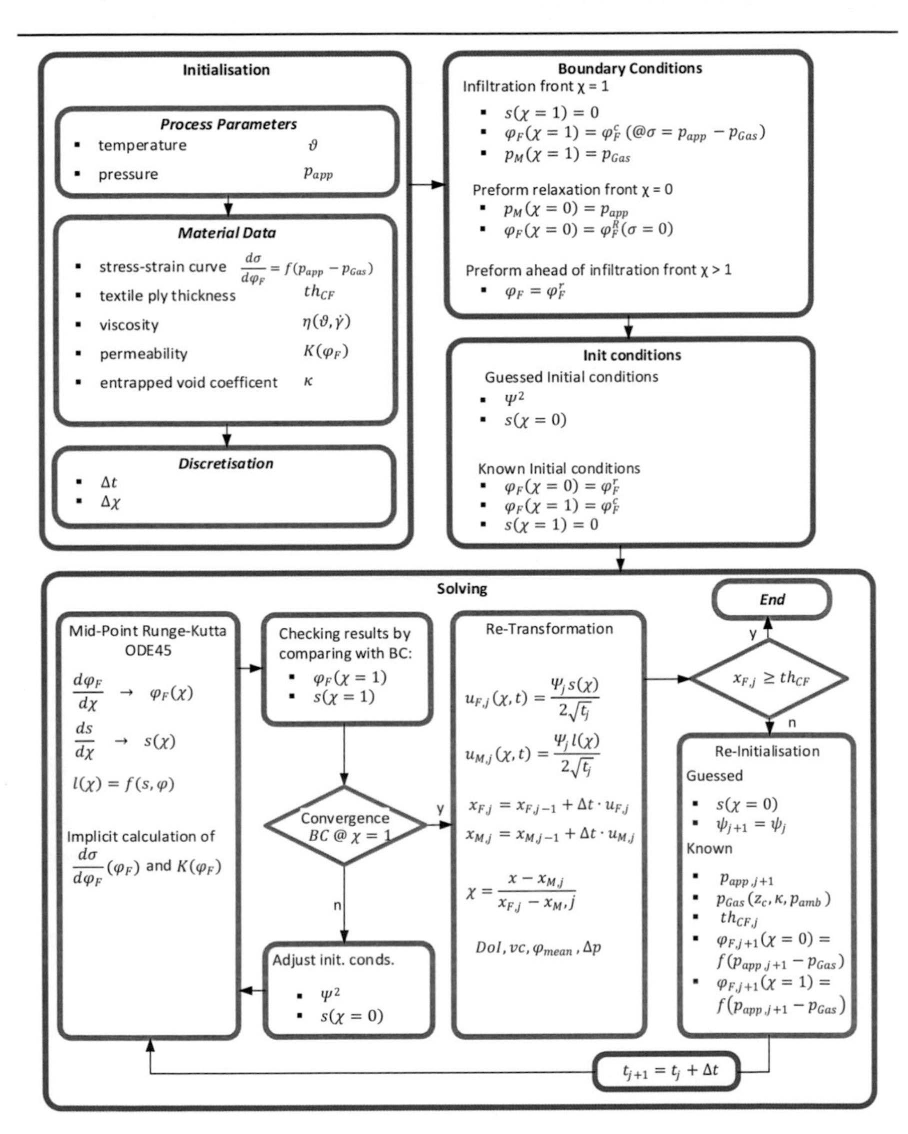

Figure 83 – Sequence of the impregnation model

List of figures

Publications and student thesis

Publications with peer-review process

Koerdt, M., Koerdt, M., Grobrüg, T., Skowronek, M., Herrmann, A.S.: Modelling and analysis of the thermal characteristic of thermoplastic composites from hybrid textiles during compression moulding, Journal of Thermoplastic Composites, 2019

Koerdt, M., Schiebel, P.: Entwicklung einer hybriden endlosfaserverstärkten Thermoplaststruktur mit integrierten Spritzgusselementen, Tagungsband des 19. Symposiums Verbundwerkstoffe und Werkstoffverbund der DGM, ISBN 978-3-00-042309-3 3-5, Karlsruhe, 2013

Publications without peer-review process

Koerdt, M., Focke, O., Herrmann, A.S.: Impregnation analysis of compression moulding of thermoplastic composites made from hybrid textiles by thickness assessment and optical analysis methods, Symposium Lightweight Design in Product Development, Zurich, 2018

Koerdt, M., Schwing, B., Bertling, H., Schreiter, M., Würtele, M., Wegner, A., Laugwitz, C.: Hybrid Structures – The Novel Way of Forming High-Performance Thermoplastic Composites for Primary Structure, ITHEC 2016, Bremen, 2016

Koerdt, M.: Developments and perspective of hybrid structures in aircraft applications, Sampe Europe 2016, Liège, 2016

Koerdt, M.: Verarbeitung hybrider Textilien zu endlosfaserverstärkten Primärstrukturbauteilen aus PEEK, Abschlussbericht des LuFo IV-2 Projekts VIA-Hybrid-FORM, Technische Informationsbibliothek Hannover, 2016

Koerdt, M., Schiebel, P., Focke, O., Herrmann, A. S.: Hybrid Textiles – The other way of forming high-performance thermoplastic composites for primary structure, ECCM17, Munich, 2016

Koerdt, M., Schiebel, P., Bostan, L., Herrmann, A.S.: Development of integral primary structures with endless fibre-reinforcement, 8. Aachen-Dresden International Textile Conference, Dresden, 2014

Koerdt, M.: Verarbeitung hybrider Textilien zu thermoplastischen Strukturen, VDI-Wissensforum, Düsseldorf, 2013

In der vorliegenden Arbeit sind Ergebnisse enthalten, die im Rahmen der Betreuung folgender studentischer Arbeiten entstanden sind

Fehrmann, Sören: Untersuchung des getakteten isothermen Thermoformprozesses zur Konsolidierung von FKV-Bauteilen aus hybriden Textilien, 2015

Grobrüg, Tobias: Untersuchung des Imprägnier- und Abkühlverhaltens thermoplastischer Faser-Kunststoff-Verbunde mithilfe faseroptischer Sensoren, 2019

Kozlik, Patrick: Prozessentwicklung zur Konsolidierung hybrider Textilien durch isotherme Werkzeuge, 2015

Mill, Simon: In-Situ Untersuchung des Verhaltens thermoplastischer Faser-Kunststoff-Verbunde während der Konsolidierung durch eine CT-gestützte Prozessüberwachung, 2018

Schings, Matthias: Untersuchung des Imprägnierverhaltens hybrider Textilien durch Messung der Bauteildicke, 2017

Skowronek, Marco: Entwicklung eines gläsernen Presswerkzeugs zur optischen Untersuchung des Imprägnierverhaltens hybrider Textilien, 2017

Bisher erschienene Bände der Reihe

Science-Report aus dem Faserinstitut Bremen

ISSN 1611-3861

1	Thomas Körwien	Konfektionstechnisches Verfahren zur Herstellung von endkonturnahen textilen Vorformlingen zur Versteifung von Schalensegmenten ISBN 978-3-8325-0130-3 40.50 EUR
2	Paul Jörn	Entwicklung eines Produktionskonzeptes für rahmenförmige CFK-Strukturen im Flugzeugbau ISBN 978-3-8325-0477-9 40.50 EUR
3	Mircea Calomfirescu	Lamb Waves for Structural Health Monitoring in Viscoelastic Composite Materials ISBN 978-3-8325-1946-9 40.50 EUR
4	Mohammad Mokbul Hossain	Plasma Technology for Deposition and Surface Modification ISBN 978-3-8325-2074-8 37.00 EUR
5	Holger Purol	Entwicklung kontinuierlicher Preformverfahren zur Herstellung gekrümmter CFK-Versteifungsprofile ISBN 978-3-8325-2844-7 45.50 EUR
6	Pierre Zahlen	Beitrag zur kostengünstigen industriellen Fertigung von haupttragenden CFK-Großkomponenten der kommerziellen Luftfahrt mittels Kernverbundbauweise in Harzinfusionstechnologie ISBN 978-3-8325-3329-8 34.50 EUR
7	Konstantin Schubert	Beitrag zur Strukturzustandsüberwachung von faserverstärkten Kunststoffen mit Lamb-Wellen unter veränderlichen Umgebungsbedingungen ISBN 978-3-8325-3529-2 53.00 EUR
8	Christian Brauner	Analysis of process-induced distortions and residual stresses of composite structures ISBN 978-3-8325-3528-5 52.00 EUR
9	Tim Berend Block	Analysis of the mechanical response of impact loaded composite sandwich structures with focus on foam core shear failure ISBN 978-3-8325-3853-8 55.00 EUR

10	Hendrik Mainka	Lignin as an Alternative Precursor for a Sustainable and Cost-Effective Carbon Fiber for the Automotive Industry ISBN 978-3-8325-3972-6 43.00 EUR
11	André Stieglitz	Temperierung von RTM Werkzeugen zur Herstellung integraler Faserverbundstrukturen unter Berücksichtigung des Einsatzes verlorener Kerne ISBN 978-3-8325-4798-1 44.00 EUR
12	Patrick Schiebel	Entwicklung von Hybrid-Preforms für belastungsgerechte CFK-Strukturen mit thermoplastischer Matrix ISBN 978-3-8325-4826-1 45.50 EUR
13	Mohamed Adli Dimassi	Analysis of the mechanical performance of pin-reinforced sandwich structures ISBN 978-3-8325-5010-3 52.00 EUR
14	Maximilian Koerdt	Development and analysis of the processing of hybrid textiles into endless-fibre-reinforced thermoplastic composites ISBN 978-3-8325-5060-8 49.50 EUR